Title of Book: The Death of Humanity - The Science
Authors: Robert Quigley & Hanan M.Tayyab

Copyright © Year 2024 Washington DC

Publisher: The Robert Quigley for President Campaign

For permission requests or inquiries, contact the publisher at:

Contact Information
Robert Quigley
142 Webster St NE,
Washington, DC 20011
Cell 202-578-8390
robertq1215@gmail.com
robertquigleyforpresident@gmail.com
robertquigleyforpresident.com

&

Hanan M.Tayyab
6812 Al Yamamah Dist,
Riyadh 12675, KSA
Cell 00966572930291
Hananhassan1820@gmail.com

Table of Contents

The new Human

Introduction

Humanity's drive to dominate nature and remake the world in our image has built a complex and voracious machine of self-supporting systems and freedom. For decades, we built a civilization with human rights, ambition, and consumption as its centered experience. This century of industrialization and technological revolutions has given us marvels from medicine to energy, transport, and communication. But underlying all this progress, a structure remains fundamentally broken. What has been created is a planet where the right to exist becomes subservient to the need to feel convenient, wherein progress outstrips sustainability and a chasm yawns between the evolution of man and survival on earth.

A Society Constituted to Expand (Growth Model)

It is the story of expansion since the Industrial Revolution. Economies are growth systems, configured to constantly grow, creating an image that failure is seen if upswing trajectories do not occur continuously. This paradigm seeps through nearly every aspect of life- from corporations, and political systems to individual goals. Nations measure success through their (Gross domestic product) GDP growth, cities relentlessly expand their boundaries and people are constantly being pushed to make more, buy more, and consume more.

Compare this to the natural world, which thrives on cycles, balance, and sustainable ecosystems. Trees grow to maturity, then level off, using resources that support life around them without ever outstripping the resources of their environment. It operates on a principle of regeneration, where the resources used are replenished over time. Humanity has developed no such principles of self-restraint or balance. Expansion has instead become our sole measure of success.

Economics and the fantasy of unlimited wealth

Contemporary economics is one of the biggest reasons for the problems of today. The global economy boils down to being an apparatus for fulfilling the desires of every individual and maximizing profit without accountability for the environmental cost of each action. Capitalism has evolved as the preferred form of economics, promoting companies to pursue higher profits in increments, creating businesses around exploiting natural resources with impunity, and dumping waste into areas without any consideration of potential long-term effects. Economists call it "externalities"—pollution, deforestation, loss of biodiversity. In reality, these are the costs of disregarding planetary limits.

Humanity's reliance on Earth's resources has become disconnected due to this economic paradigm. People who live in cities thousands of kilometers from the woods, seas, and rural areas are unaware of the damage and exploitation taking place to support their way of life. This disconnection fosters a sense of security that feels false, the earth appears endless and resources are never-endingly abundant. With climate change proceeding at warp speed and ecosystems crumbling, the flaws in this economic design become embarrassingly apparent.

Individual Fulfillment and Democracy

Democracy is the source of liberty and sovereignty, where self-asserting and personal decision-making are at its core. However, it stands for sacrifice in the long run for collective needs rather than short-term satisfaction. Democratic systems rely on election cycles, which force elected leaders to try to solve short-term problems that will make them popular rather than working toward a long-term consensus that will give health to the planet. Measures that would change the environment and reform proposals are unpopular because they call for tough sacrifices voters do not want to make.

Democracy has in so many ways proven a means for protecting and expanding individual liberty in the name of environmental responsibility. The delicate balance of freedom and collective action to survive remains difficult to achieve. Climate change, resource depletion, and other planetary emergencies put democratic societies to their greatest test. Anything that was envisioned for such a system was designed more to respond to immediate insights not community sacrifice or long-term ones survival goals.

The Life Span Illusion and Humanity's Blind Spot

Most humans live only 70 to 80 years, which is a short period in the timeline of Earth. This very short time defines for us, in turn, the sense of urgency and disaster. A world seemingly boundless with almost limitless resources creates a world of security that is more illusion than reality. In every generation, collapse happens to the next one - or at least, each believes so. It is hard to fathom the urgency and magnitude of the current environmental crisis when the effects take decades and centuries to be evident.

This means each generation passes the burden of finding a solution to the next one. Rather than sacrificing today for a tomorrow far beyond our lifetimes, there's an expectation that the technological or policy breakthroughs that will be made tomorrow will take care of the problems we leave behind. Yet with each passing generation operating in this pattern, the edge draws closer, with less and less to mitigate its effects.

A growing population and unsustainable demands on the planet

The unchecked population growth has increased the strain on Earth's resources. Since the 19th century, the world's population has grown rapidly, resulting in congested cities, higher demands for food, energy, and land, as well as an increase in trash. Earth can comfortably support about 2.5 billion folks. With over eight billion people vying for limited resources, the ecological footprint of humanity has grown much beyond what the earth can sustain. Every new head causes added pressures, but all technological inventions can slightly extend available resources rather than counter and control exploding populations.

From forests cut down to create farmlands to the draining of water sources, greenhouse gases choking the atmosphere to each extra individual being that one more consumer in an already too-consumed system is an added force to the call for dramatic changes in the way humans distribute and consume resources.

The Rope and the Noose

Humanity's technological power created for us at once the tools of our salvation and the mechanisms of our downfall. The pathways of renewable energy, innovations in sustainable agriculture, and emergence in carbon capture technologies offer ways toward mitigating some of the worst effects of climate change; but these technologies cannot, of their power, fundamentally change underlying structures and mindsets that drive unsustainable behaviors. The same technology that can pull us back from the brink also has led to unprecedented resource exploitation and environmental degradation.

Shared Sacrifice and Delayed Gratification

It will take an entirely new paradigm, based on sustainability, shared sacrifice, and long-term thinking, to avoid environmental collapse. The goal is to shift the mindset towards a culture that prioritizes planetary health over individual gain by delaying gratification and sharing sacrifices. In order to replenish resources, consume less, and provide for the needs of tomorrow, human systems must operate within limits.

This would question practically every aspect of modern living. It would challenge our thinking, our dependence on growth to measure success, require restructured economies that reward sustainability, and reset personal consumption choices. The shared sacrifice in this regard would be almost painful but is the only way forward for a species that has nearly exhausted its natural inheritance.

The Hope for a Symbiotic Future

It has been an age of discovery, innovation, and expansion, in the full flush of forming a world unprecedented in history. Yet during this age we unfolded the cost of unchecked and unbridled progress causing environmental ruin, resource depletion, and created precarious future prospects. Indeed, it was the rope by which such a civilization as the current one can lift itself from the brink of extinction only to become that noose that tightens around a future.

As this world comes to realize the limitations of its growth and the fragility of its systems, it stands at a crossroads: the possibilities for sustainable, symbiotic, harmonious coexistence with nature are finally within reach, and only if this society can, at last, dismantle the pillars of limitless expansion, instant gratification, and lonely individualism. And so perhaps by embracing a future grounded in equilibrium and interdependence, humanity might yet discover a way to survive as humbly respectful of the Earth- not as a resource to be conquered but as a partner in survival.

Chapter I

The Edge of Existence

The world wasn't going down without a fight, though it was dying. The earth warned the humans daily in the winds as it howled, from the sun's burn and silence of where life once had been abundant. Those who could hear it recognized the echo that haunted them daily, a chilling hint that its end was no longer far off but an imminent fact drawing nearer with every inhalation.

Dr. Amina Hassan, one of the top climate scientists at NASA, stared out longingly over a circular bay window. For years, she had made a clarion call for the looming catastrophe and after all this time meticulously examining these environmental changes that were moving humanity closer to its tipping point. She looked again towards the far smog clouds, feeling guilt sting her conscience. Had she done enough? Could she have done more? She pulled herself away from the window, surrounded by clammy faces that sat huddled together around the coffee table.

Michael O'Connor, an Irish politician who had fought a losing battle against the forces of corruption and denial lamented: 'It's incredible how low we have sunk.' He glanced at Amina with defeat in his voice, hoping for some solace. We put in place policies, to slow the harm down, but never enough. The greed, the apathy… it just seems so pointless at this time.

Sofia Morales, an incredible Mexican activist and journalist nodded in agreement. For years, she had been out on the front lines of it — showing us what climate change looks like and how much it is hurting. Everywhere I go, it's the same old thing, she spat "Droughts, wildfires, floods… it feels like the earth is at last striking back & there's nothing we can do about that"

The portly Indian tech entrepreneur Arjun Patel stood there, his arms hugging across his chest. He knew that everything he had trusted in, technology, and more importantly the conviction "that all his saving of endangered species could save us", was starting to fall apart. "Here, we thought if we just innovated our way out of this," his voice gone bitter. "Yet we have made the situation worse. The more we make, the worse that things become. That is a terrible spiral, and I do not know how we can avoid falling into it.

The other person who sat still beside Arjun was his wife Clara Santos, a Portuguese artist, she has always been a dreamer, having faith in art and its ability to move mountains. That belief was being eroded, did you ever think about the future? She whispered, her voice quavering, longing for offspring, "What it would be like to bring a child into this world"? I had always wanted a family, but now… I don't think I can handle it.

At Clara's words, Amina felt a twinge of sadness. Her daughter Leila sleeping in the next room, blissful of what her world would be like. Amina said in a quiet voice, "I think about it all the time." whenever I looked at Leila, I asked myself" Did I make a mistake? She sighed ….. Hmm. Then again, quitting is not a choice. For ourselves, for our children, and for the world….

"We just need to keep trying".

This time Sofia was beyond frustrated, she shook her head. "But how? It is so deep. It is almost rooted in everything we do. What could we do that would make a difference?

The question weighed in the humid air as a heavy reminder of their powerlessness. They were nothing but a pack of friends, trying to navigate in an unraveling world.

The conversation had grown intense. Almost boiling over with the weight of the issues they faced.

They spoke of the vanishing rivers and lakes, of farmlands that had become too barren to grow anything, and the choking effect of deforestation that was slowly suffocating the planet.

"We have all heard the stories," Clara murmured, her voice heavy with sorrow, "of people being driven from their homes, forced to cross into wastelands just to survive."

A deep silence filled the room. The despair was thick, almost suffocating.

It was late, and outside, the city had faded into a distant blur, a shadow of what it once was. The streets, once full of life, now felt haunted, echoing with the memories of better days.

Amina broke the silence, her thoughts still churning from everything they had discussed. The anxiety, the frustration, the sheer terror of it all.

It was almost too much to take in. But she knew, deep down, that they couldn't afford to lose hope.

"We do not have all the answers," Amina said quietly, but with a strength in her voice that steadied the room. "But we can't stop now. We have to keep trying, keep searching for a way through this. It's the only chance we have got."

The others nodded slowly, determination settling into their features. They understood how daunting this was, how nearly impossible it seemed. But giving up wasn't an option. They had to keep fighting for themselves, for the people they cared about, and for a future that still seemed worth saving.

When they finally parted that night, there was a sense of renewed purpose among them. Each carried the weight of what lay ahead, but also a fierce resolve to make a difference, to fight for a world that was slipping away. It was a huge challenge,

Chapter II

Water's Last Whisper

The heat of the morning seemed to be all too typical, and the thick morning air felt especially so. With her heart heavy from the knowledge she held, Amina Hassan went out into the pale light. The sun, which was once a representation of growth and life, now loomed down like an overseer, its beams cutting through the haze with fierce cruelty.

She snapped out of her reverie when her phone buzzed in her pocket. It was a message from Dr Ethan Rhodes hydrologist, her best friend and colleague whom she had worked with for many years.

Although his remarks were succinct, the urgency in them was clear: "We must speak. The information is concerning!!"

Though Amina had prepared for the worst when she went to see him.

The world was running out of time and options.

Ethan waited for her at one of the last few cafes that still tried to feel like the old days. His face was drawn, the lines around his eyes deeper than she remembered.

He wasted no time with pleasantries.

"Amina, it's worse than we thought," he said, his voice low.

"The Ogallala Aquifer is on its last legs."

Here is the report summary. Ethan quickly opened the file,

The USGS report (2023) indicates that water-level changes in individual wells from predevelopment to 2019 ranged from an increase of 86 feet to a decrease of 265 feet. On average, across the aquifer, there was a decline of 16.5 feet during that period, with a slight rise of 0.1 feet from 2017 to 2019. The change in recoverable water storage was a decrease of 286.4 million acre-feet from predevelopment to 2019, but there was a small increase of 1.6 million acre-feet from 2017 to 2019. In 2019, the aquifer held approximately 2.91 billion acre-feet of recoverable water. These measurements were taken from 2,741 wells.

We've got less than sixty years before it's completely dry. She felt her breath catch in her throat. Sixty years? Ethan, that's impossible. We were supposed to have more time. He shook his head, the weariness in his eyes unmistakable. "The droughts are getting longer, the extraction rates are higher than ever. It's not just the Ogallala, every major water source we've been monitoring is in critical condition. The Colorado River, the Indus, and the Yellow River they're all on the brink."

"Take a look at this," He pulled another file.

According to projections, around 700 million people could be displaced due to severe water scarcity by 2030, and by 2040, approximately 1 in 4 children globally will live in areas experiencing extremely high water stress (UNICEF, 2021).

He looked dispirited and said, "If current conditions persist, the world may exhaust its water resources even before sixty years.

He showed her another report. "Look Amina the suffering is about to begin"

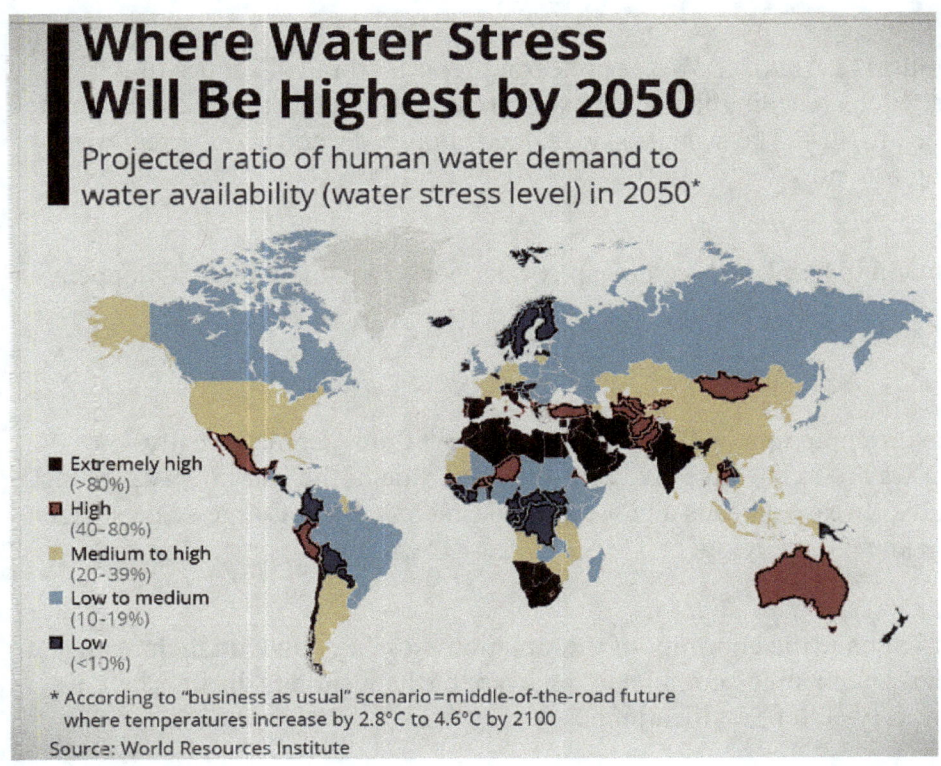

Sourced by World Resources Institute

Amina looked at him, her mind was unable to fathom what he was saying. "What about desalination? Is there any way we can scale up fast enough?"

Ethan sighed and ran his hand through his graying hair. "Desalination is a Band-Aid, not a cure. It's energy-intensive, and the environmental toll is enormous. We're talking about billions of people who rely on freshwater sources that are vanishing. Desalination isn't going to save them."

Their conversation was interrupted by the arrival of Sofia Morales, fresh from another harrowing assignment. Her face was gaunt, her eyes hollowed by what she had witnessed.

"I just got back from South America," she said, her voice raspy. "The Amazon's tributaries are drying up. Whole ecosystems are collapsing and the people there. They're desperate. They've started to move north in droves but there's nowhere to go. The land is dying".

Ethan nodded gravely.

Sofia continued, "It's not just the Amazon. I've been hearing reports from colleagues in Africa and the Middle East. The Nile is receding faster than was predicted, and the Tigris and Euphrates are in places all but reduced to dust. The scenarios we talked about a decade ago the real worst-case ones? They're coming true now".

Amina felt a cold sweat begin to break out on the nape of her neck. We knew it could happen but not so soon. Not like this.

Sofia leaned in" her voice dropping to a whisper. "Amina, I don't think people actually understand just how close we are to the edge. Water is only one side of the story it's everything. Food production is crashing, disease spreads and we see the first real sign of large-scale society collapse. The world as we knew it, is dying".

They were sitting in heavy silence the enormity of the situation pressing down on them. Outside, the city buzzed with its daily happenings completely unaware of the insidious disaster being unleashed just beneath the very skin of civilization.

"We really must do something", Ethan finally whispered.

"But what? The governments are locked, industries are over-involved, and the public either denies the fact or is too burdened to act". Amina's mind busied furiously in search of a way out but all it could find were dead ends.

The problems were too wide-ranging and interlinked. They had spent their lives trying to warn the world yet now that the crisis was here they were powerless to stop it.

"We can't just give up," Sofia said with fierce determination on her face. "There has to be something we can do".

Vague thoughts raced through her head, as Amina nodded slowly. "We must find a path to the people who have the power to change things. We have to move beyond the science, beyond the statistics. We have to make them see what's coming."

Ethan looked at her with a small ray of hope, which had flickered anew in his dull and tired eyes. "What are you thinking about?"

Amina wavered as she knew the risks involved. "We need to show them the reality not just in abstract terms, but in a way that they can't deny. We have to make them feel the gravity of this matter, the fear, the desperation. We have to make them feel that it is now or never. This is our only chance to act".

"How do we do that?" Sofia was frowning.

Amina's gaze drifted to the window where the gray sky looked as if it would explode. "I have an idea," she said slowly. "But it's dangerous. For us to really make an impact here, we have to get the truth out unfiltered, raw, and terrifying. We have to show them what life is going to be like when there will be no more water."

Ethan and Sofia looked at each other in worry. "You mean... you want us to show them the worst-case scenario?" Sofia said trembling a little bit.

To that, Amina nodded. "Yes. We bring them to the places where water has already gone, where life is just on the brink. We show them what the end of the world looks like." A deep, uneasy silence fell over the table. Although it was a desperate plan, it was also their last hope.

"If we do this," Ethan said softly, "there would be no point of return. We could lose everything our careers, our reputations, and all you can think of."

Amina looked him straight in the eyes, "If we don't, we lose the world."

Sofia took a deep breath; her quivering hands were her only tell. "I'm in," she said calmly, the anxiety in her eyes betraying her emotions. "I don't care what it will take. I'm in."
"All right, let's do it." Ethan's face looked determined once again as he nodded. "Let's show the world the ugly truth they've been trying to look away from."

As the trio left the cafe the weight of what they just set in motion fell on their shoulders. The road was dangerous and unclear but there was no turning back now.

In the gathering darkness, Amina felt the first electrifying sense of something she hadn't felt in a long time, suspense, an unsettling sense of foreboding that gripped her heart. They were on the verge of crossing a line that arguably might change everything. She walked into the night with the feeling of being watched and that something or someone was waiting at the edge of existence.

Chapter III

Fading Fields, The Vanishing Green

"This land killed me."

The words were scribbled hastily, almost as if they were too painful to write. Sarah's small trembling hands clutched the piece of paper her wide brown eyes tracing the jagged lines over and over, trying to understand. She was only thirteen a child who had just returned from school expecting nothing more than a warm meal and a tired but loving smile from her father. Instead she was holding the last words he would ever write.

A knot of worry began to tighten her stomach. Something was wrong she could feel it-----, even if she didn't fully understand what was happening. Clutching the note tightly, she wandered through the house, her eyes scanning each room as she called out for her father.

"Dad?" she whispered, her voice trembling with fear. There was no answer. The house, usually filled with the sounds of life, was quiet.

She checked the kitchen, then the living room moving from room to room her heart pounding louder with each step. She was holding the note in her hand as if it were the only thing tethering her to reality. But no matter where she looked, she couldn't find him.

Finally, she made her way to the barn. The door was slightly ajar creaking as it swayed in the breeze. For a moment........ She pushed it open......

There hanging from the rafters was her father.

Sarah froze, her breath catching in her throat as she stared at his lifeless body swaying gently in the still air. The world around her seemed to tilt, everything slowing down as the reality of what she was seeing downed on her. The note fell from her hand to the ground … FORGOTTEN…

She wanted to scream, to cry out for help but no sound came. She was paralyzed unable to look away, unable to move. She felt like sinking in the horror of the scene. Her strong, kind, loving father was gone. And she was left standing there a terrified little girl alone in the world with nothing but his final words to haunt her for the rest of her life.

A soft knock on the door brought Sarah back to the present. She realized that she had been staring at the spot on her desk for who knows how long, lost in the past.

"Miss Hayes, are you still there?" a colleague's voice cut through the fog of her memories. The concern in their tone was clear. You should go home it's getting late.

Sarah looked up startled as if seeing the lab for the first time. The sterile bright lights and the hum of equipment seemed almost foreign, after falling so deep into her thoughts. She nodded quickly trying to shake off the lingering sadness.

"Yes, I was about to leave," she replied, her voice slightly hoarse. She forced a small smile, though it didn't quite reach her eyes. "Just lost track of time."

Her colleague hesitated for a moment, watching her carefully, then nodded and stepped out, leaving her alone once again.

Sarah sighed and began to gather her things, her movements slow and deliberate. The drawer where she had placed the note remained closed, but she could still feel its presence, a heavy reminder of the past.

For Jack Hayes, the suffering had started so small and slight you would hardly know it was there — almost haphazardly with a drop in water supply that once flowed daily to irrigate his farm. He had relied on the wells and a nearby lake to irrigate his crops for years, bountifully nurturing them in much the same way as his father and grandfather before him. However, as the seasons went on deepening wells dried up to contain only small amounts of water and the lake diminished gradually where its surface became worryingly shallower than it used to be.

Jack, fearing for his dying lands began referring to all types of wisdom he could find. He took it upon himself to familiarize himself with the facts on the rate at which the water table is being depleted through excessive pumping of water from boreholes etc. Late at night he would be reading about the impacts of over-extraction of water from the ground. He did not doubt that the larger picture became far more daunting as he expanded his knowledge of the situation. On one of these evenings, Jack happened to see a map showing how the global water table was dropping.

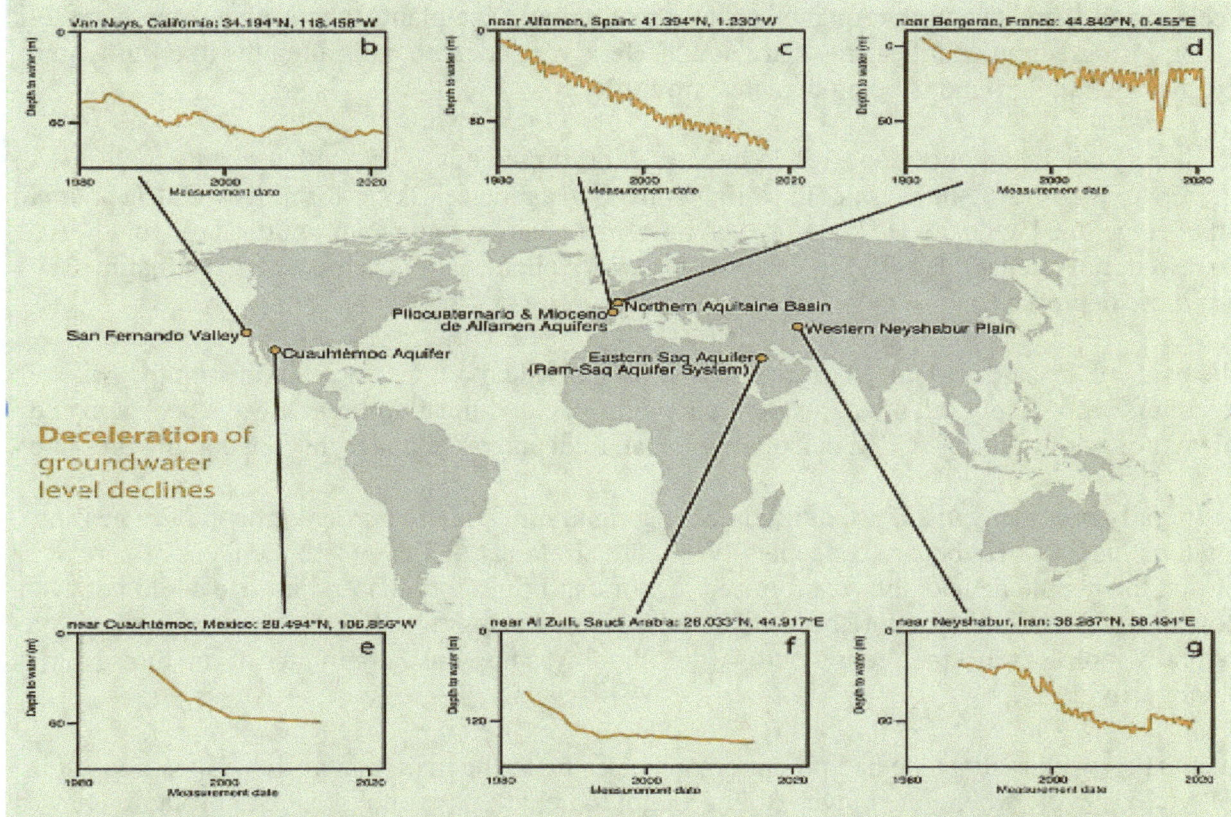

Illustrative examples of individual monitoring wells that record cases for which groundwater levels declined during late twentieth century and continued to decline but at a slower rate in the early twenty-first century (that is, decelerated deepening) a, Global map depicting the locations of the six monitoring wells (that is, each point represents one monitoring well). The aquifer system that each monitoring well lies in is labelled next to each point. b–g, Measured groundwater-level variations over time for individual monitoring wells. Each panel presents groundwater-level data for a single monitoring well.

Sourced by European Commission 2024

This made him understand that worldwide water tables were declining at a worrying rate, a calamity that humans had caused, due to too much pumping of water and at a rate that was beyond that which nature could resupply. New wells would have to be dug even deeper meaning a lot more money than he was capable of at present.

But digging deeper required not only resolve but also cash - much more cash than Jack had ever had to raise. He went to the bank, with a heavy heart explaining that he was to buy new equipment that could reach the lowered water level. He hoped with all the passion of his soul that when he got this loan there would be a shot at this farm saving project of his. The loan was arranged but with strings attached; one which Jack was willing to meet since he thought he could bring change.

That is why during the next few months, the situation started to change, and the hope of staying at work and improving one's status gradually weakened. Then the new equipment was procured and the wells were deepened further, but the water they got was still less. There was no rain for weeks on end, and the earth remained as barren as before. The plants that used to grow were full of life, lush green in countless rows are now lifeless; reminders that his attempts to outgrow the rate at which the ground is being depleted are futile.

The bank, which had once appeared to be the savior, turned out to be another torture to deal with. The first notices came in ……..cold calling collection agencies asking for money that Jack could not even get his hands on. He had not even been in a position to pay a proportion of what he had borrowed; not even a quarter of it. Every letter, every phone call, became a new cut that made him more depressed.

The bankers, he said, did not even treat him as an individual but just as a number on their list of defaulters who simply did not care about his plight. Jack who had hitherto always been a man of few words with a strong-willed personality began to disintegrate under this unbearable pressure.

He literally saw the walls of the building caving in on him and the burden of his failure bearing down on his back. He began to sleeplessly wander around at nights, where the faces of his family, the people he had always promised to work and fight for, chased him. He could not even look at his wife's face anymore because he failed her as a husband. His daughter whom he used to fondly look forward to greeting daily after school was now just a reminder of the future that he cannot provide.

Eventually, Jack realized he had no option, hence came to the main conclusion. He is trapped in that vicious cycle where one moves from drought to debt to despair and back to drought. He was a man who had been reduced to a state where he could be pieced back together and he was worn out by forces that were outside his realm.

A notion hit Jack like a hammer as he strolled across the barren fields that had once been lush and alive. *THE LIFE INSURANCE POLICY…...* Years before, when things were more promising and the farm's prosperity was certain, he had taken it out. It was a tiny solace he had stashed away for his family, a safety net intended for a distant and far-off moment.

But now that his failures were bearing down on him, that approach appeared to be the only viable course of action. With stark clarity, he recognized that his death may finally put an end to the financial issue that was casting a storm over his wife and daughter. He felt a deep, gut-wrenching grief as well as a twisted feeling of comfort at the concept.

Maybe this was the only way he could support them right now, he thought angrily, metaphorically the last crop produced on this desolate farm.

Homer's death, Jack specifically, was brutal and brought to the attention of the community the sad reality of the losses they incurred due to the environmental and economic disaster that had consumed their life. He was buried, but people also gave vent to fears—a symbol of the greater evil that is befalling the dying society around them.

Years passed, but the pain never truly went away. It was always there, lurking in the back of her mind, a constant reminder of what she had lost. But it also fueled her, driving her forward in her studies. She had to understand what had happened, why the land had failed them, and why her father had felt so hopeless.

She threw herself into agronomy, determined to find a way to prevent what had happened to her family from happening to others. She studied the soil, the weather patterns, and the impact of climate change on agriculture. She learned about the new technologies being developed, and about the efforts to create crops that could survive in even the harshest conditions.

Now Dr. Sarah Hayes was no longer the scared little girl who had found her father's body in the barn. She was a scientist, a woman who had worked so hard to ensure that no other family would have to suffer the way hers had.

Her research was more than just a career it was a mission. She was determined to create crops that could survive in the kind of conditions that had killed her father's dreams. She wanted to give farmers like him a fighting chance.

As she examined the seedlings under the microscope, she allowed herself a moment of hope. The plants were thriving, even in the simulated conditions she had created in the lab. They were resilient just like she needed them to be. But the real test would come when they were planted in the field, exposed to the elements and to the same forces that had ravaged her family's farm all those years ago.

Sarah knew that the road was still long. If she could succeed if these crops could truly withstand the changing climate then maybe she could find some peace. Maybe she could finally lay her father's memory to rest.

She stood up and stretched her tired muscles. As she gazed out of the window at the city, she thought of her father of the fields that had once been green now long gone.

I'm doing this for you Dad......, she whispered her voice barely audible in the empty room. I'm going to make sure your sacrifice wasn't in vain.

Sarah turned back to her work knowing that there was still so much to do. But she wasn't alone in this fight. She carried her father's memory. The only way to honor his sacrifice was by ensuring that the fields would be green again.

Chapter IV:

Breath of the Atmosphere

The nutty aroma of fresh coffee in the warm air classroom made it easy to forget the freezing Siberian surroundings. Rahim sat near the window, his notebook open but not paying attention. Showing a diagrammatic representation of the global warming chain.

Here you can see the process of global warming, filling up entire space with his words was Professor Aleksandr Dimitri's Voice

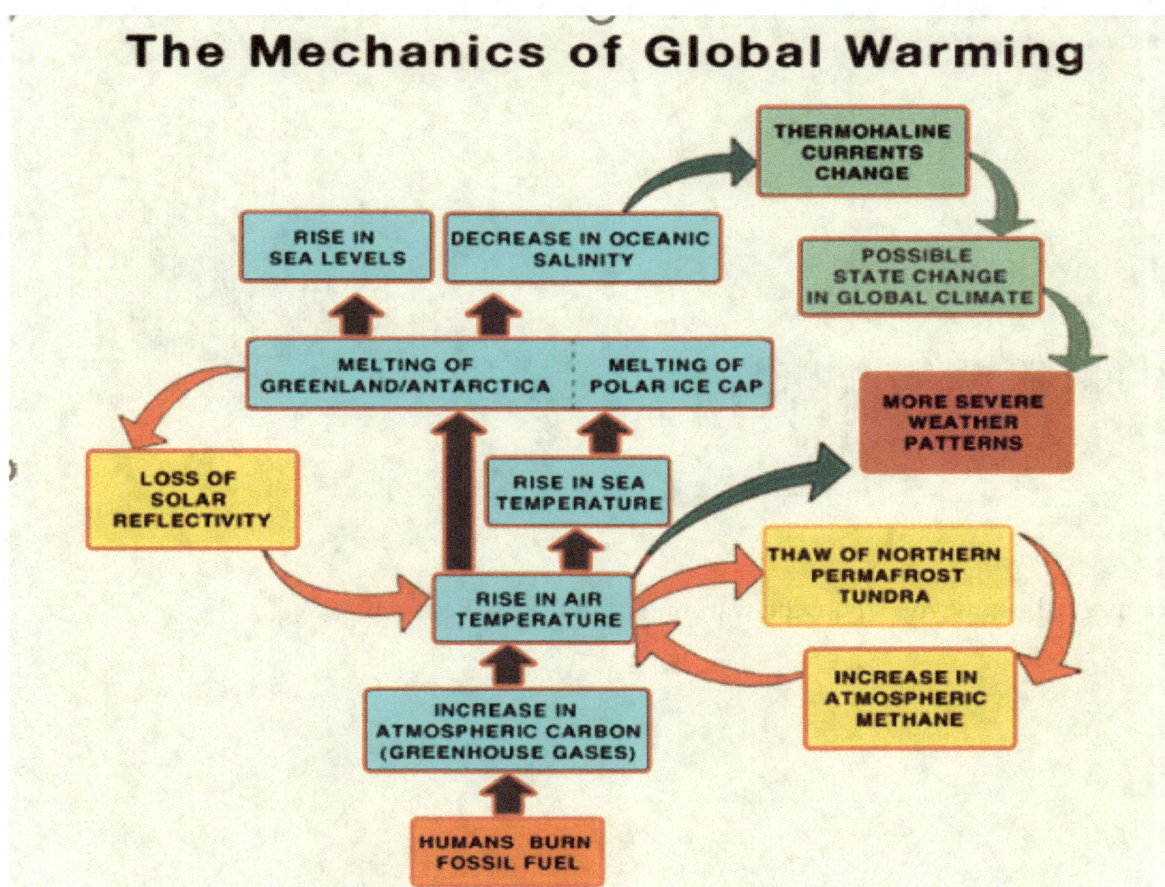

The Mechanics of Global Warming

Sourced by thehcf.org

He added "In the diagram on the right, the central vertical stack of blue boxes represents the main chain of global warming events, with fossil fuel emissions at the bottom. When the Arctic and Greenland ice sheets melt and the planet loses its vital solar reflectivity, we get the feedback loop on the left (red arrows). The feedback loop that occurs as the frozen arctic tundra thaws and releases methane, a powerful greenhouse gas, is shown on the right (red arrows)".

Professor paused a little, then he asked, what is climate change?

Robi answered, "Climate change is the phrase used to describe long-term variations in temperature and weather patterns. Significant volcanic eruptions or variations in the sun's activity might be the source of these swings. But since the 1800s, human activity—primarily the burning of fossil fuels like coal, oil, and gas—has been the main driver of climate change.

As a result of the burning of fossil fuels, greenhouse gases are released into the atmosphere, enveloping the earth like a blanket and trapping solar heat, raising global temperatures.

Very good, he continued, "In this graph, you can see the rise in temperature throughout the past century"

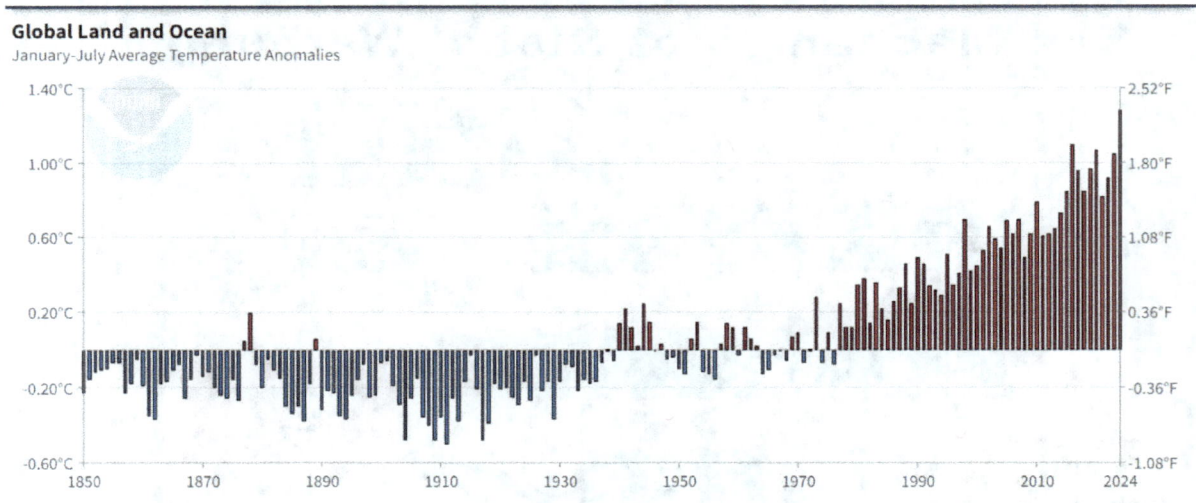

Global Land and Ocean

January-July Average Temperature Anomalies

"This picture will show you where the warming of the earth the is worst."

Land & Ocean Temperature Departure from Average Jan-Jul 2024
(with respect to a 1991-2020 base period)
Data Source: NOAAGlobalTemp v6.0.0-20240806

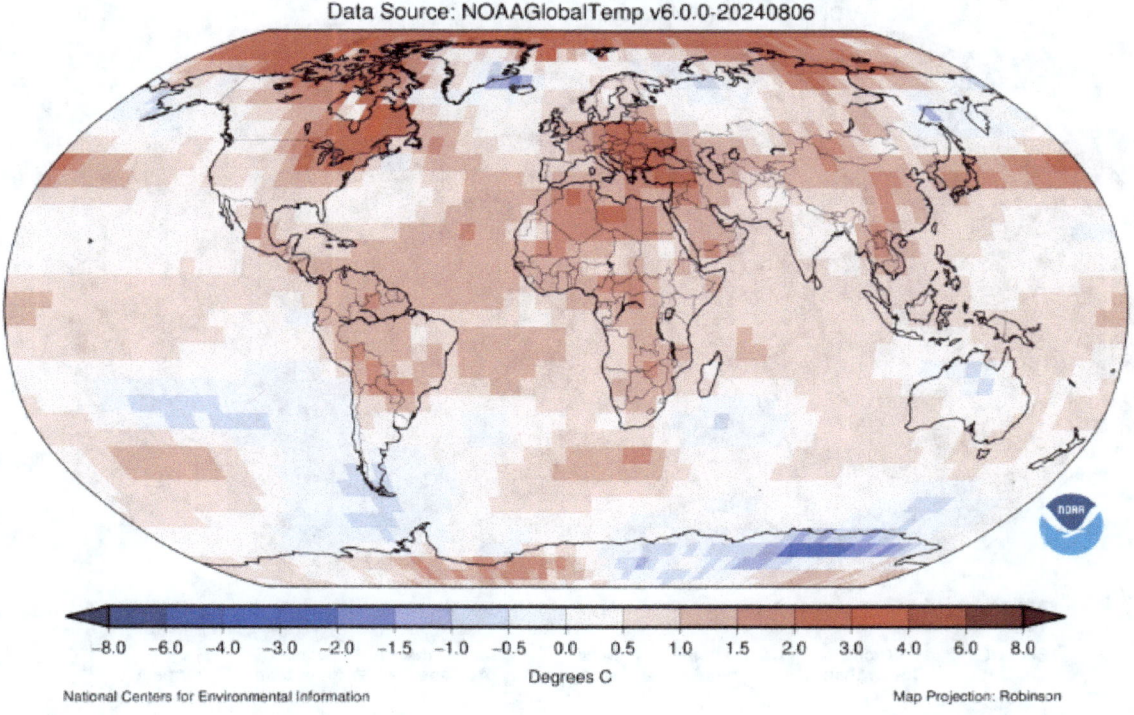

National Centers for Environmental Information

Map Projection: Robinson

[January–July 2024 Blended Land and Sea Surface Temperature Anomalies in degrees Celsius](#)

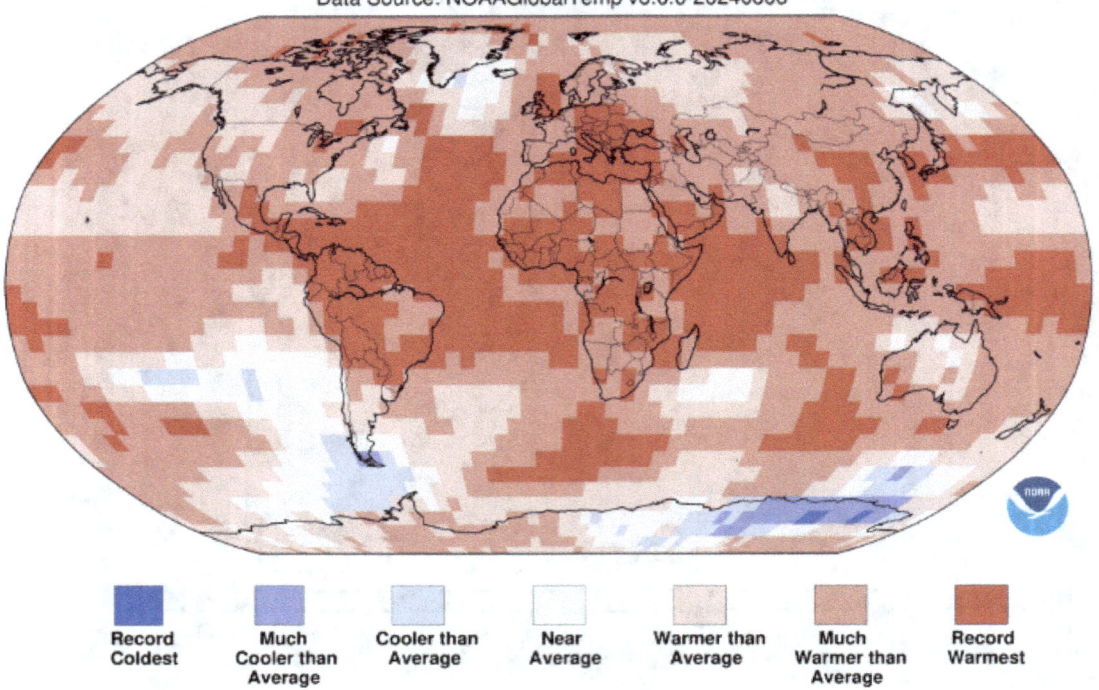

Land & Ocean Temperature Percentiles Jan-Jul 2024
NOAA's National Centers for Environmental Information
Data Source: NOAAGlobalTemp v6.0.0-20240806

| Record Coldest | Much Cooler than Average | Cooler than Average | Near Average | Warmer than Average | Much Warmer than Average | Record Warmest |

January–July 2024 Blended Land and Sea Surface Temperature Percentiles

"Record-warm temperatures were for the January–July year-to-date period across most of the northern two-thirds of South America, as well as in much of Central America and Mexico. Parts of southern Asia, much of Europe, and much of Africa have seen record-warm temperatures so far this year. Much of the Arctic, as well as portions of eastern Canada and the northeastern United States, had temperatures between January and July that were more than 2 to 3°C (3.5 to 5°F) above average. A large portion of Greenland, Scandinavia, central Asia, and the southern and western United States were also unusually warm. Eastern Antarctica saw much cooler than average to record frigid January–July weather, making it an atypical mild spell. Additionally, there were areas of eastern Greenland, the southern point of South America, and portions of northern Australia, and parts of northwest Russia and the Russian Far East that were extreme.

The majority of the equatorial Atlantic and a sizable portion of the subtropical Atlantic had record-warm sea surface temperatures over the first seven months of this year, following a pattern akin to prior year-to-date periods. A large portion of the southern Atlantic, southern ocean, western equatorial Pacific, and the Indian Ocean were all experiencing record temperatures. Other portions of the world saw abnormally warm sea surface temperatures, except for the southwest Atlantic and southwest Indian Ocean, the tiny portion of the North Atlantic between Greenland and Iceland, and sections of the southeastern Pacific and Southern Ocean.

Some areas of the western Pacific saw sea surface temperatures that were somewhat below normal".

As the Professor surveyed the room, his eyes fell on Rahim, who sat unusually silent. Rahim was always quick to answer, his hand often the first to rise with insightful responses. But today, there was a distant look in his eyes, his usual attentiveness replaced by something else. The professor hesitated, almost calling on him, but decided to wait. Perhaps it was just a rare moment of distraction. He would ask Rahim about it after the lecture.

Returning to the lecture, the Professor said, "The permafrost thawing is one of the consequences of global warming. It not only comes with weather changes it also releases age old pathogens," It is a little more direct, but a somewhat less obvious explanation that climate change includes deadly impacts on human populations — especially on the people who have lived near the ice for so long."

The lecturer continued, "Every year as the permafrost is melting … we are seeing changes in climate and historical pathogens coming back. This is a real, but somewhat obscure story that climate change does not only mean warmer temperatures — it also affects the human adaptations and communities living near ice for thousands of years.

Rahim was trying to focus when he noticed a strange sensation throughout his body. He noticed that his hands were trembling a little instead of being as steady as they usually were. The warmth of the room, which had been comforting at first, was oppressive now. He looked around, watching as everyone listened closely to the lecture, their faces wearing satisfied grins. His stomach suddenly went to mush at the scent of coffee, how tantalizing.

The warmth of the classroom was abruptly replaced with a rush of cold. His eyesight became blurry due to his intense shaking. In an attempt to find some comfort, he reached for his coffee cup but accidentally knocked it over. Rahim lost consciousness as soon as the cup hit the floor.

A wave of panic swept through the classroom as Rahim's other classmates came to his aid. The professor called for help and minutes later an ambulance was on its way. Rahim got too cold. It looked like he went into shock and passed out.

They wheeled him into the hospital and dozens of tests were conducted as they tried to determine why he had just gone down. His friends and teachers waited, hours turned into days of worry. The diagnosis, after all that time, would simply be unimaginable. Rahim had anthrax disease that hadn't surfaced in over 100 years and was the first case diagnosed in modern times. *What a coincidence** he thought bitterly…

Doctors said the source of Bacillus anthracis spores, which lay dormant for centuries under ice, before thawing permafrost released them back into freshwater on the Yamal Peninsula. The disease — which had previously been limited to ancient times, was beginning to surface in the reindeer herds, and even more frighteningly, among human beings. The outbreak served as a cruel reminder that uncertainty did not only apply to the future but also extended far into history.

Hospitalized, Rahim lay still and considered his whereabouts concerning the rest of those he loved—half a hemisphere away. There was something he wanted to tell them, but it would not come out, memories still rose uninvited in his mind from the list of their ordeals. He remembered how his parents, Areef and Laila — now penniless at home in India after using their savings to ensure a better life for him and his sister, Aisha. Their path had neither been short nor smooth.

Once, Areef and Laila had lived a pastoral life following the slow rolling tides of the sea. The Maldives, a paradise of turquoise waters and powdery white sands, had been their home, where time seemed to come unstuck between the never-ending blue sky above and the dark azure sea below. The life here was easy but very enriching. They were from the long line of a family firm based on commercial fishing, whose craft had been passed down through several generations and still operated using ancient tradecraft dating back centuries. The sea was their best friend, an endless blessing feeding them not only in body but also in spirit.

For everything else, the power was nature in the Maldives. Palm trees lazily shook in the hot, salty winds, and below they found life filled the tops of coral reefs. The islands were a chain of beautiful atolls, one more lovely than the next, and from all appearances far removed from life's traumas. Each day the sun rose and set over waters of gold, reflecting here and there where the waves broke in rows along a silent stretch of shore. Each night music drifted through moonlit air as fingers plucked strings while abominations sang softly to oceanic tides.

However, this tranquil life was not going to last.

Over the years, however, their family business had become a dicey proposition as the ocean just east of Uruguay turned against them. The waters were so bountiful before they started to invade their land. At first, the changes were modest, a touch less shore here and tide just somewhat higher there. But it quickly became apparent, that water was inexorably rising. The fish —their lifeblood— started disappearing from the reefs, chased off by warming waters and an acidifying ocean. The fish ceased ever spawning in what had been their dominion. The island to which they owed their entire heritage, their bedrock with centuries of history and tradition was being eaten up by the oceans that gave life.

Their reality was the distant concept of climate change that scientists and news channels regularly talked about. At that moment, sea-level rise was not a theoretical concern but an existential one.

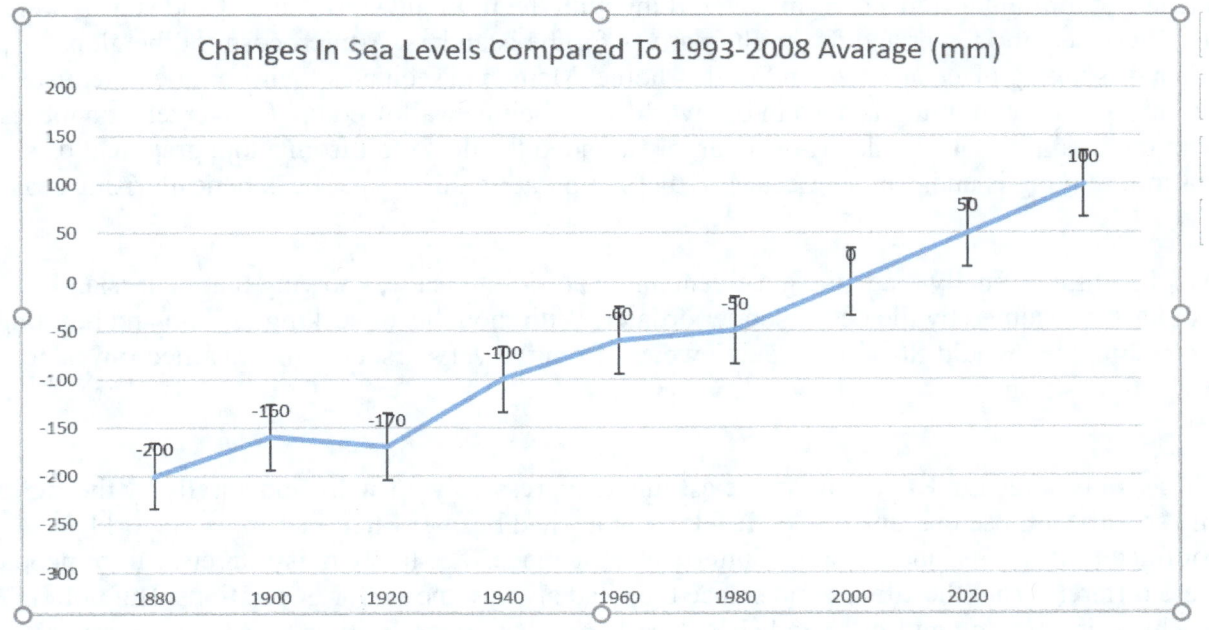

The Maldives is especially exposed due to its low-lying atolls. The ocean stole further each year, eating the shore side stone and salt to earth and water.

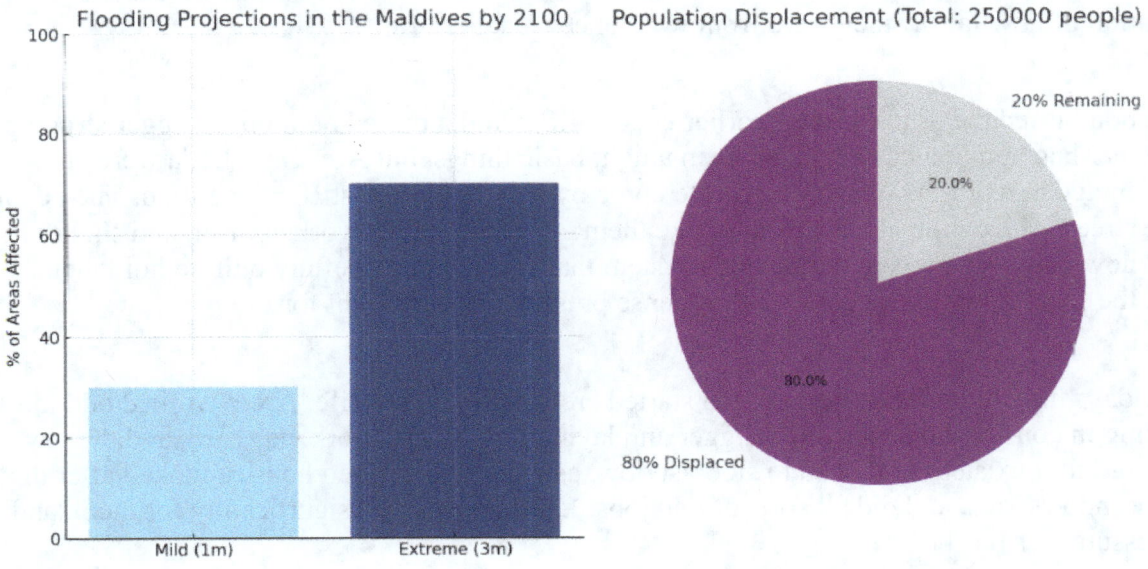

To compound the problem, extreme weather events increased in number and intensity. Storms, as yet so few and far between still resembled gentle assaults upon the islands but they grew in number. The monsoon raged in overtime fury, the rain turned to rivers, and houses were drowned and swept through torrents carrying everything with them. Rapid storm surges and storms were relentless, leaving fewer and fewer alternatives for the islanders. A microcosm of the global crisis was taking place at Areef and Laila's home. Many had seen past their imminent demise, and most of the low-lying islands in our world were being swallowed up. Conversely, countries with mountainous terrain like Taiwan largely avoided the death toll from a topographical clash, but in low-lying islands among oceans of the South Pacific, they sank back beneath rising seas.

Areef and Laila finally made the gut-wrenching decision to leave, knowing that their island would be consumed by the ocean sooner or later. With thoughts of seeking asylum and building a new life, they would flock to India. However, from the very first day, they learned it was no perfect transition.

These souls landed in Kochi only to be assaulted aggressively, visually and aurally by the sights and sounds of those bustling colors. It felt like they had been airlifted into India, a world confined to its chaos, and put in this utterly chaotic place. The lively noisy streets, full of people were different from the silence and solace they had always known for generations. But not in Kochi — its teeming multitude and chaos nearly ran them over. Being strangers in a strange land, from the get-go they faced innumerable difficulties.

The language was the first and most immediate obstacle. In the Maldives, they spoke Dhivehi, a language not at all related to any tongues spoken in India. Hindi, and Malayalam — words they were unfamiliar with. Unable to communicate or integrate, they largely became isolated – unable to work or find connections in their new country. Real loneliness kicked in, as well as the realization of how far off they were from what once was their life.

They soon buried the ashes of their earlier era, as life in India turned out to be a bigger struggle than either had anticipated. They had been taught basic things, but Areef and Laila were at a disadvantage in real life. Work was hard to come by and they faced stiff competition. The people on the streets of Kochi had been exactly like them — searching for work, a meal, and shelter in an unknown city. They were trapped in a system that seemed purposefully built to hold them down, the sheer number of people put immense pressure on their fight for life.

As the days turned to weeks, desperation started creeping into their life. Areef worked odd jobs, including in construction, as a ditch digger and laying bricks. The work was hard, and the pay was poor but it existed. Laila had to stoop so low, as well. She cleaned houses, looked after the neighborhood's kids, and did all sorts of odd jobs. Life had changed significantly for them, and the pressure was on.

Then for Areef and Laila, instead of languishing in past pain – shaped by this thing that happened to them, they wanted a bit more bite. That their children would have it better. At their expense, they gave it their all for Rahim and Aisha to go to school. Despite their age, the children understood clearly how much their parents had sacrificed for them. Getting good grades, and fighting to get one step ahead while coming from a background where the education system was not as beneficial in some ways.

Years of perseverance slowly paid off for this family. Rahim whose father had left him with a keen mind for knowledge was lucky enough to have gotten a scholarship to study in Siberia.

For Areef and Laila, it was a moment of unbearably sweet happiness — relief that their hard labor had been proved correct. But life in Kochi continued to drag them down even as they cheered for Rahim's success.

Kochi, with its smog-filled air, compared to the clean and fresh air of their hometown Maldives. Laila's health worsened gradually. The hacking cough she had suffered with for years began to worsen until, finally, she received a diagnosis of lung cancer. It read like a cruel irony — here was someone who had once breathed the cleanest air on earth, slowly succumbing to Kochi's toxic atmosphere.

But as Laila deteriorated, her husband faced a choice between looking after his bed-bound wife and providing for their family. He was weighted by their circumstances, but resolute. They had made it through so many hardships that they would make it past this as well. It was a high price but it was purchased for her that she had to fight every day against the relentless reaping of poverty, disease, and hope.

Areef and Laila were one story among countless others about a planet tipping over the edge of climate change with pressures further exacerbated by starving population demands. Their travels were set against the backdrop of a disaster that not only threatened their family but countless others across the world. Communities were rising, driven from their home, decimating livelihoods, and shifting around the world essentially seeking a safe place.

Even with the tremendous odds stacked against them, Areef and Laila held onto this hope that their children would rise to own a world that another generation had sacrificed so much for an unborn future. That small hope they clung desperately to died a little with each faint echo from the sad world around them.

The news had been devastating, but she hid it from Rahim so that he could concentrate on his studies.

Rahim came back out of his thoughts, in a hospital bed somewhere in Siberia. He knew his sickness would stress out his parents, but he also remembered the struggle they went through to bring him up. They had fought and he would fight as they did. Somehow their sacrifices lent him the courage to confront whatever nature might present.

The fact that anthrax had broken out—and more broadly, that diseases once thought eradicated might rise again—from deep history was a wake-up call to the capricious future human beings could expect. Yet Rahim understood his family's narrative was lived strength, confront all problems. They had been through the worst and seemed to have made it.....

Now it's his turn.

When he recovered he went back to India, where he longed to be with his family. But when he arrived, the shock greeted him — He learned that his mother had been diagnosed with lung cancer. The years she spent living in Kochi breathing in the polluted air severely affected her. This realization hit Rahim like the wave and he started to lose control over his mind. Thinking once of going to Siberia with his family turned into his firm decision. Finally, he could stand it no longer, he must get them out. Siberian air is cool and clean. Rahim swore that he would take them with him to the far north.

Rahim had been planning this process for several months. His mother's nagging cough which was a constant reminder to them of the years of hardship only worsened each day. The once energetic big city now appeared oppressive, its population density of millions giving bodies their rhythm. Thus, the pressure of people, fighting for territories, food, and air to breathe became unbearable. All this in India where the population has been rapidly increasing and the air getting thick with pollution made him and his family feel like being enclosed in a box.

Rahim started to convince his father to relocate. He saw his scholarship as a way out to a country that most wouldn't think of, Siberia. It was a sight of a future that he felt was drawing closer. The peripheral zones of the Russian empire where the air was thin, the territory was vast, and people scarce appealed to him. Where Kochi was pushing, condensing, bodies and buildings against one another, Siberia was opening, fresh, calm and full of potential.

Siberia, once hidden deep under ice, was turning green and opening up its vast planes, heated for thousands of years. Here things changed. The Earth was growing, awakening to a new day. This place, which had been a cold and barren wasteland years ago was gradually turning into a place that could offer hope to a world that was growing increasingly hot and crowded. Few comfortably habitable places remained on Earth. This was where Rahim helped his family feel like they could start a new life from scratch.

With this picture in mind, he decided on the risky move of taking his family to Siberia. It was not a comfortable discussion that they had. His father Areef continued to encourage himself that things could probably improve in India. However, what was heading toward Rahim was no secret to him, he had the figures, and the prognoses. The facts were not hidden from him.

That, however, was not the end of being a benefactor in the eyes of Rahim's vision of his family. He started to contact his friends in India as long as they were under the same dilemma too. "Follow us," he encouraged them to do so. He understood that many of them had the same challenge of eradicating inadequate resources, scarce air, and a challenging life where everything seemed to be in short supply. Siberia, he told them was the future.

In the past, people never considered Siberia a place they could live. Yet society was evolving. Places around and adjacent to the equator, which had once supported human and animal life, started to become desert-like. Food shortage and famine were only a few months ahead due to droughts, heat waves, and failing crops affecting millions. Rahim realized that it was only a matter of time before the twenty-first-century humans would have to run to the cooler latitudes in search of inhabitable lands. Siberia is one of the few areas of the earth that has not been fully developed, due to the huge resources in the region. This wasn't beyond irony – what used to be deemed as far too cold, even uninhabitable, was perhaps the best bet for life to exist.

With the shrinking of the permafrost, the land revealed itself with all its glory – fertile soils that when frozen could not support living organisms. Finally, the forests started to expand their area towards the north, free interacting rivers started to exist, and the wild animals came back to the area. However, this was accompanied by new hazards which were never a threat in the past due to scarcity. The thawing ground also released ancient dangers: The ice where the bacteria and other pathogens had been encased for thousands of years suddenly started to wake up. This led to the release of methane gas which is regarded as rather damaging to the green climate of the Earth and, paradoxically, hastened the very climate change that made these northern tracts inhabitable.

Despite this, it went deep into the heart of what made the adventure appealing as Rahim knew all these dangers. Humanity had little choice. That is why in the middle regions the earth was becoming uninhabitable and many people would be forced to look for space near the poles. The higher latitude areas such as Siberia will become the new frontier of man's civilization. It was no more a question of survival of man; it was about ensuring that the piece of this upper middle land thaws, grows green, and is the first to be claimed by them.

In Rahim's mind, he was a trailblazer, one of the few who understood that the destiny of the world was in these northern lands. The narrowly cramped streets of Kochi and the pressuring population density of India were just the start of a new world change. As the Earth's climate persisted to warm, the majority of the population would be forced to migrate, constrained by the required living conditions into those regions characterized by cooler climates and more hospitable landscapes.

The experience of his family was a sort of rehearsal of sorts for what happened later on. Rahim had his future in a sword that was pointing towards him but at the same time, he was a bearer of all things that were to come, where man was to shift from south to North and change the course of mankind in the future.

Chapter V

A Perilous Inheritance

Rahim had always thought that his life was just a thread in a vast tapestry. He had grown up on the fragile, vulnerable islands of the Maldives, where the daily rising tide was an apt reminder of how easy it was for everything to be swallowed by the sea, and he had soon realized that life's comfort is a fragile thing. Now, he stands on the frozen horizon of this vast expanse of frozen tundra, stretched out with a strange sense of relief, as if he would find an answer somewhere here, in this land of endless snow and stone, to all his questions.

It was more than just survival that he wanted. Rahim was looking for a reason, an excuse for the unstoppable momentum of a world that had gained speed for nearly two hundred years, feeding on itself and stretching out to every corner of the planet. The students he spoke with in Siberia—a mix of Russians, a few South Asians like him, even some Canadians and Americans—each carried their own pieces of the world's chaos. Many longed for a fresh start, an untouched sanctuary where they could somehow make a break from the weights of the past.

Walking along to the university library on this winter morning with the sun low on the horizon, he finally met with Professor Ivanov, a history teacher with whom he had become involved. Ivanov was going through the unintended consequences that technological progress had on civilizations in the course of history. It seemed to Rahim that Professor Ivanov could see a thread of this modern fight linking back to the rising and falling of societies ages ago. Moving through stacks of ancient texts and modern texts, Rahim was aware of how deep the professor's thoughts were.

"Tell me, Rahim," Ivanov said with a low voice, against the hush of the library, "do you know what it was that doomed civilizations before us?"

Rahim thought about it. "War, disease. The collapse of resources?"

"Ah, yes," Ivanov nodded, a faint smile tugging at his lips. "But more than anything, it was expansion. An inability to see limits. They pushed outward until they ran out of land, resources, and even time. And then, without the wisdom to change course, they simply. stopped."

The words of the professor struck him like a revelation, echoing deep into his soul, sensing something growing in the Maldives, where the boundaries of existence were so palpable and immediate.

Here we are, Ivanov continued, believing we've conquered nature, that our technologies and industries make us invincible. Yet, for 175 years, every decision we made, every law, and every invention, was founded on the assumption that there would always be more land, water, and freedom to do as one pleases. Now, they stand at the edge.

The whole time Rahim had only suspected the presence of this yearning for "more"; it hadn't dawned on him that this weakness was a legacy, created right at the start of industrial society. Since then, ever since those initial factories went up and their smokestacks darkened the sky, humans organized the society with growth in its core. They made the economy into a religion that told the people that good was when they had many things and consumed them well, and that expansion became all justification for any expenditure.

"People thought of the economy as something separate from the Earth itself," Ivanov said, eyes piercing through the dim light. "But in fact, it's nothing other than a function of resources of the Earth. Those that don't grow overnight."

This was a society that, in creating it, had both empowered and imprisoned it. It gave humanity the greatest tool: the heart of the Sciences and the science of everything. But it also locked human society into a paradigm where the Earth was something to use and throw away rather than to love.

Professor Ivanov handed Rahim the newspaper, his eyes heavy with sorrow. "Read this article," he murmured, voice laced with despair. "See for yourself... how devastatingly global warming is tearing our planet apart."

A feedback loop: what is it?

"Processes that can either amplify or diminish the effects of climate forcing" are known as climate feedback loops. ("Forcing" refers to the primary factors that influence our climate, such as solar radiation, greenhouse gas emissions, and airborne particles that originate from both natural and man-made sources that affect our climate, such as dust, smoke, and soot.)

Simply said, feedback loops intensify or diminish the effects of major climatic causes, causing a cyclical chain reaction that keeps happening.

Positive and negative climate feedback loops are the two main types for our needs.

A process known as negative feedback reduces function, frequently in an attempt to stabilize the system.

On the other hand, a positive feedback loop "accelerates a response."
In the case of the water vapor cycle, it runs somewhat like this:

1. The atmosphere warms as a result of increasing emissions of heat-trapping greenhouse gasses.
2. More water evaporates from our lakes, rivers, seas, and land and enters the atmosphere as a result of this warmer air.
3. More water vapor is held in warmer air, and water vapor retains heat.
4. The initial warming is increased because the additional water vapor in the already heated air holds significantly more heat.
5. The cycle is restarted when further warmth causes more water to evaporate. And once more. And once more.
It's a vicious cycle whereby climate change triggers a series of consequences that lead to further climate change. An issue that we caused is taking on a life of its own, with possibly disastrous results.

A positive loop can eventually get out of control without the negative feedback loop's regulating function, changing the climate system in ways that are irreversible. We refer to this as a "tipping point."

NOAA states that irreversible tipping points—that is, climatic changes that are not stable and predictable—can jeopardize the increasing impacts of positive feedback loops. Tipping points are essentially little alterations in the climate system that have the power to transform a rather stable system into a completely other state. When a wine glass tips over, for example, wine spills out of it and cannot be recovered by standing it up; the glass's previous condition of being full changes to one of being empty.
All of this means that even if we haven't reached a tipping point yet, many aspects of our climate system are already contributing to harmful positive feedback loops, which exacerbate the effects on people worldwide and compound climatic conditions.

Climate and Feedback Loops
There are now two striking instances of positive climatic feedback loops occurring in the Arctic. The first is occurring on land, where the climate crisis is causing permafrost, which contains significant amounts of carbon and methane, to thaw. The second is on the Open Ocean and ice.

One of the most potent greenhouse gases is methane. Methane is very short-lived in the atmosphere compared to CO_2; in 20 years, just around 20% of the methane released today will still be there. Nevertheless, it traps heat around 120 times more effectively than CO_2 when it initially enters the atmosphere and 86 times more effectively during a 20-year period. (CO_2 persists for a very long time; in 10,000 years, up to 15% of today's CO_2 will still be present in the atmosphere.)

Melting Ice, Warmer Earth

The decrease in sea ice cover, especially during the summer, is another Arctic positive feedback loop with worldwide ramifications.

The amount of ice in the Arctic Ocean is essential for controlling marine and land temperatures worldwide. A vast white surface formed by sea ice reflects sunlight away from the planet. This is called "albedo," and sea ice excels at it in comparison to other earth surfaces.

Regretfully, the Arctic's sea ice cover is decreasing. In terms of the second-lowest minimum on record, Arctic sea ice extent this past summer was statistically tied with 2007 and 2016. (The ice extent was the lowest on record in 2012.)

"The ice extent has decreased by 40 percent since 1979, and this loss is changing Alaska's climate, increasing coastal erosion, decreasing habitat for walruses and other marine mammals, altering the timing and location of blooms of the microscopic plant life that makes up the food web, and decreasing survival rates for young walleye Pollock—the largest commercial fishery in the country," according to NOAA.

More ice loss results from this depletion of sea ice, which also fuels additional climate change and global warming. (Dominoes dropping, remember?)

This is due to the fact that the much darker ocean surface below is visible when sea ice is absent. Additionally, unlike ice, which reflects sunlight back into space, open water absorbs it.

How much radiation from the sun are we discussing here? The open ocean absorbs the remaining solar energy and warms the water and surrounding atmosphere, reflecting just 6% of it back to space. On the other hand, 50–70% of the solar energy that enters the earth is reflected by sea ice. A colder surface and a functioning climate system are the results of less heat energy being absorbed.

An irreversible tipping point is rapidly approaching in the melting of Arctic sea ice. In fact, many scientists think that the question of when we will start experiencing summers in the Arctic without sea ice is more important than if it will happen.

According to Scientific American, "we may get our first glimpses of ice-free Septembers in the next 20 to 25 years, when we will have added another 800 billion metric tons of CO_2 to the atmosphere," given the present world emission rates of 35 to 40 billion metric tons per year. However, it doesn't end there. With more CO_2 in the atmosphere, other months of the year will have no ice. For instance, the Arctic is expected to lose its ice from July through October if an additional 1,800 billion metric tons of CO_2 are released.

This evaluation, it should be noted, comes from an article that acknowledges "observed ice loss is generally happening faster than climate models have forecasted" at the beginning.

Sourced by
The Climate Reality Project. (2023)

Rahim stared at the article, the words blurring as his mind drifted into a sea of doubts. Was bringing his family to Siberia truly the safe haven he'd hoped for? He'd convinced himself this was their best chance, away from the relentless effects of climate change back in India. But as he read about the warming earth, the shifting climates, and Siberia's own uncertain future, he couldn't shake the gnawing worry that he might have uprooted them in vain.

In another corner of the university, Rahim met Emil, a German climate science student sharing his passion for sustainability and technology. Emil was sketching out a project comparing disparate energy sources: wind, solar, geothermal, and even nuclear. They spoke about nuclear power in its "big" and smaller, newer modular reactors that Emil argued could save lives if responsibly developed. Still, every form of energy came with its own compromise.

"It's like a rope," Emil said, and he quickly sketched out a rough design on the chalkboard in the study room. "Our technology could lift us out of the pit we've dug ourselves into, but the same rope can hang us if we're not careful."

But what stuck in Rahim's mind was the rope. Dually symbolic of hope and peril, it stood in for the human impulse: industrial might had lifted him, but it pushed the planet to breaking. It was clearer than ever now that their ways of thought—those systems of "limitless" economics, politics, and individualism—were founded on the presumption that resources and safety were infinite.

With Emil's maps in hand, Rahim envisioned his native Maldives—a paradise slipping below the waves, a cautionary tale of both nature's power and humanity's stubborn will to dominate it.

Yet here too, the problem stuck: how were they going to renew without reliving the mistakes of old? He began to consider a new model in which bounds are not transcended but, rather, mark out a line in the sand to be respected. A world in which sustainability wouldn't be something opted for but the only option, one with sacrifice and waiting inside a people that would, instead of fleeting self-advantage, lay a civilization upon it.

As winter deepened, Rahim started to feel that people on campus were waking up to this vision. There had to be a way of living that didn't operate according to the growth paradigm but according to the rhythms of the Earth rather than always against them. It was a model inspired by the Earth's rhythms, stable, renewable, and unassuming.

One night, by the common room fire, they started to discuss what they were going to do next. Some wanted to turn the icy landscapes of Siberia into sustainable villages, others wanted to start rewilding forests in parts of Europe, while others aimed at rebuilding desert areas into resilient agricultural havens. They shared their ideas regarding energy systems, food production, and how to be in harmony with Earth's processes. It felt as if something new was unfolding, tender and tentative as the hope that arose with the understanding that things could never have been so heavy and wrong in the past.

Chapter VI

The Last Conference

The rain drummed incessantly on the old university building outside like an endless war where a quarter was not given. Inside, over the murmur of machines, the air was tense. This had once been the center of excellence in academia; it was now a haven for the last hope of humanity. In the dying heart of a sinking city, the best minds that the world had ever seen had gathered one last time to report on their progress and confess fears.

The scientists sat in a long row around the table in what had once been a lecture hall. A flickering overhead light cast sharp shadows on their tired faces. Outside, all was desperation; cities were about to fall, crops were dying, and millions were dead either due to civil war *or* sectarian conflict or resource scarcity and these few, however, had not yet given up.

From the head of the table, Dr. Elena Kovic, a veteran atmospheric scientist and former colleague of Dr. Amina, cleared her throat as if asking for silence from the rest of the team.

"We're getting somewhere," she said with a calm voice but weakly. "But come on, let's just level with each other here. Some of these solutions, most of them- they're just not cut out for it. Let's just go over it together and what is working, and what isn't, in your mind."

A brief murmur ran through the room as each scientist steeled themselves to confront the hard truths they'd been shying away from.

The Carbon Trap

In 2050, you leave the Permian Basin Petroleum Museum in Midland, Texas, and drive north across the sunbaked scrub where a few surviving oil pump jacks sit idly in the heat, and then you'll see it: a glittering palace rising out of the pancake-flat ground. The land is mirrored here, with the choppy silver-blue waves of a massive solar array stretching out in all directions and lapping at a massive grey wall that is five stories high and nearly a kilometer long. Behind the wall, you can see the curving pipes and gantries of a chemical plant, "Energy expert Raj Patel saying imaginatively".

Carbon Engineering is planning the world's largest direct air capture plant, in Texas, USA (Credit: Carbon Engineering)

Showing the image, he continued, the wall appears to be moving and shimmering as you approach; totally composed of enormous fans spinning within steel boxes. It appears to be an enormous air conditioner that has been blown up to unbelievable proportions, you think. This direct air capture (DAC) facility is one of tens of thousands that exist all over the world. They are working together to remove carbon dioxide from the atmosphere in an effort to chill the planet. During the 20th century, billions of barrels of oil were extracted from the depths of this

Texas terrain, making it renowned. The CO2 in our air, which is the legacy of those fossil fuels, is currently being poured back into the depleted reserves.

Explaining in detail Raj proceeds, direct air capture is a simple scientific process. The method used by Carbon Engineering's system uses fans to pull air that contains 0.04% CO2 (today's atmospheric levels) through a filter that is soaked in potassium hydroxide solution, also referred to as potash, a caustic substance used in soap making and other applications. After the CO2 in the air is absorbed by the potash, the liquid is pumped to a second chamber and combined with calcium hydroxide, often known as builder's lime. Little flakes of limestone are created as the lime absorbs the dissolved CO2. In a third chamber known as a calciminer, these limestone flakes are sieved out and heated until they break down, releasing pure CO2, which is then collected and stored. The remaining chemical leftovers are recycled back into the process at each step, creating a closed reaction that never runs out of resources.

But take a minute to travel back in time to 2021, to Squamish, British Columbia, where a barn-sized gadget shrouded in blue tarpaulin is being finished against a picturesque backdrop of snow-capped mountains. Carbon Engineering's direct air capture plant prototype will start removing one tone of CO2 from the air annually in September. This is a tiny start, and there are plans to build a somewhat larger facility in Texas, but this is the standard size of a DAC plant nowadays.

Leaned forward, tapping fingers drumming restlessly on the tabletop. "DAC is pulling Co2 from the atmosphere but barely at the scale we need. It's like trying to empty the ocean with a spoon."

"The energy demand is a problem," Elena said, nodding her head. "We've tied it to the solar grid, but even with all the sun we're getting, the power output just isn't stable. We need geothermal, or we need a decentralized grid."

"That's the problem," said Raj, shaking his head. "Geothermal plants are too time-consuming to construct and not always feasible in a location. If we get decentralized energy, electric vehicles and home solar systems will still not be enough to run DAC at all times. And let's not forget all this is occurring at the height of social collapse. People aren't exactly lining up to ditch their cars and houses for the sake of saving the planet."

And I don't want to forget mentioning the cost …. 'Raj added'.

Here's the direct air capture cost in USD for every ton of CO2, as per CO2 concentrations.

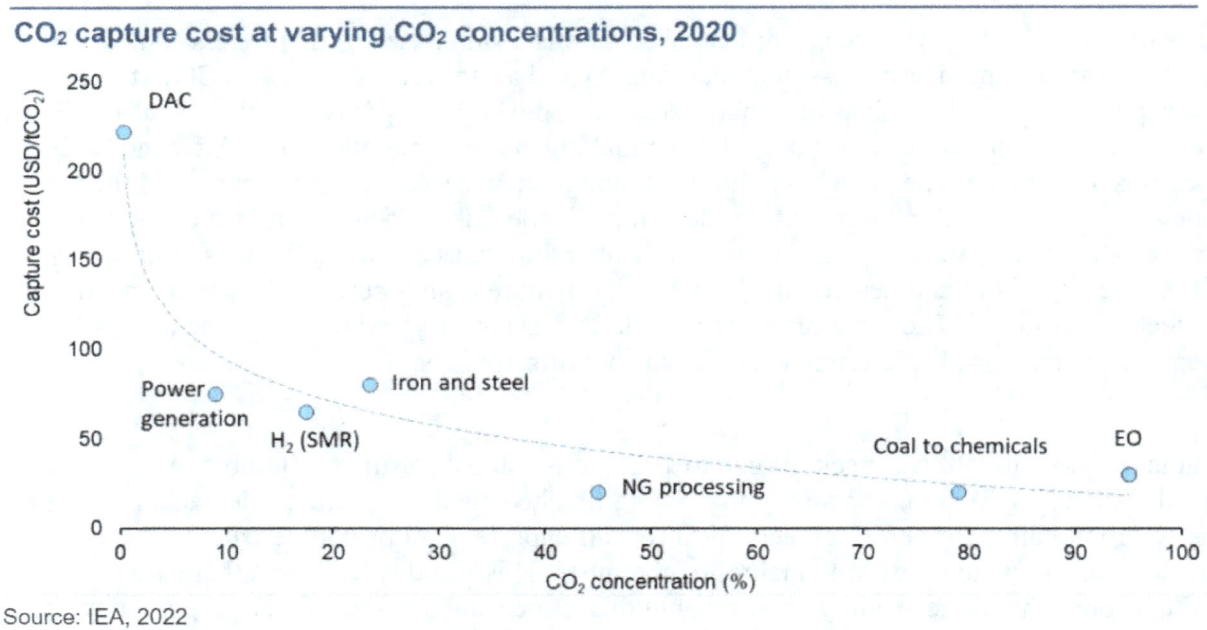

CO₂ capture cost at varying CO₂ concentrations, 2020

Source: IEA, 2022

The room was silent. Dr. Kovic turned to her colleague, Mei Tan, who was in charge of the research on bioplastics and microbial solutions.

The Plastic-Eating Army

Mei took a deep breath and pulled out a report in front of her, and started explaining.

A member of the Comamonadaceae family and genus Ideonella, Ideonella sakaiensis is a bacterium that can break down and consume polyethylene terephthalate (PET), utilizing it as a source of carbon and energy. A silt sample collected outside of a plastic bottle recycling plant in Sakai City, Japan, served as the initial source of the bacteria.

Ideonella sakaiensis is a rod-shaped, aerobic, gram-negative bacteria. Cells have a single flagellum and are mobile. I. sakaiensis colonies are round, smooth, and colorless. Its dimensions range from 1.2 to 1.5 μm in length and 0.6 to 0.8 μm in breadth.

Here is the picture for better understanding.......

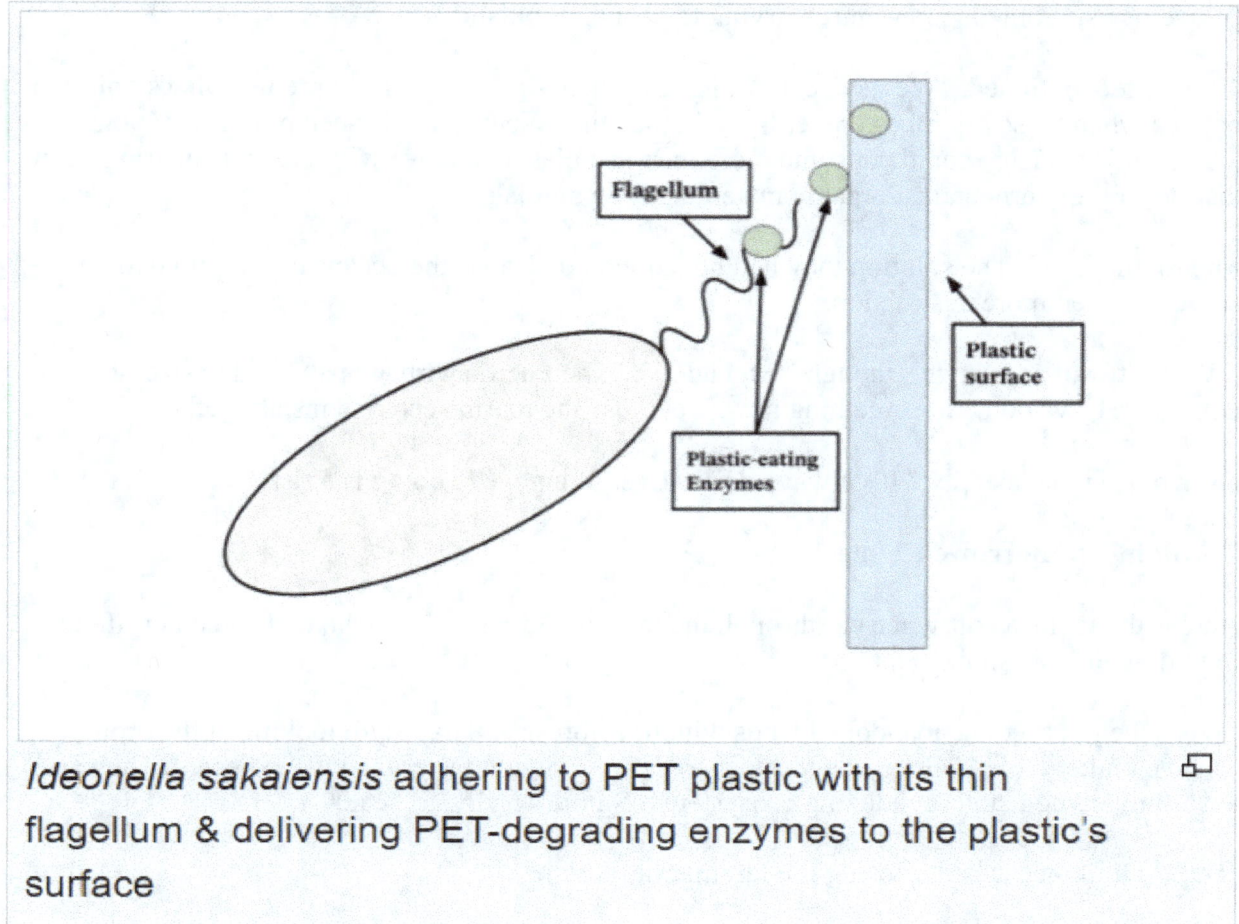

Ideonella sakaiensis adhering to PET plastic with its thin flagellum & delivering PET-degrading enzymes to the plastic's surface

Image taken from Wikipedia

She continued, the ability of Ideonella sakaiensis to degrade PET in fisheries that are fed sewage is being investigated. It has been demonstrated that different strains of this bacteria do not provide any risks to fish development and culture. This bacterial species effectively uses PET as a carbon source and flourishes in wastewater and aquatic habitats contaminated by plastic, demonstrating its potential as an affordable anti-pollutant.

Since Ideonella bacteria are incredibly sluggish, they are not very useful in the battle against plastic waste. The bacteria took an astounding six weeks to break down a 20 mm long and 15 mm broad piece of plastic sheet.

In order to break down PET more quickly, the enzyme PETase, which breaks down PET plastic, has been genetically engineered and coupled with Monohydroxyethyl terephthalate (MHETase). PETase also breaks down PEF (polyethylene furanoate) polymers. This and further strategies might be helpful for recycling and upcycling mixed plastics.

"Our microbes are promising. We have engineered them to break down most plastics, especially the stuff polluting our oceans. They're eating through it faster than anything we've seen before.

"Good news," someone murmured, trying to sound optimistic.

"But," Mei continued, "We have control issues. There's no way to fully predict what's going to happen when these microbes are let loose in the wild. We've already seen minor mutations under lab conditions. They could consume more than just plastic, and we will end up triggering a new disaster if they turn against organic materials, even animals."

The room tensed. The solution they had hoped would cleanse the oceans might prove to be the source of even more destruction.

"We can't wait any longer, though," Mei added. "The Pacific garbage patch is the size of a continent. If we don't do something now, we'll lose the marine ecosystems altogether."

Dr. Kovic sighed deeply. "It's always a risk. Everything we're doing is a risk."

Rewilding on Borrowed Time

Across the table, a soft voice cut through the tension. Sarah Hayes, who dedicated her life to rewilding efforts, spoke next.

"We've begun seeding abandoned lands with new forests," Sarah said, looking at the drone footage on her tablet. "Gene-edited super-trees are growing faster than we expected. They are absorbing carbon, and revitalizing ecosystems. Nature's trying to heal."

"But?" Dr. Kovic asked, sensing the familiar hesitation.

"The problem is time," Sarah said. "Forests take decades, even these modified ones, to reach full carbon-sequestering potential. We simply don't have decades. Our models show that if we'd started this process thirty years ago, we might have had a chance. But now? These forests will grow, yes, but they won't stop the impending collapse."

She paused, then added, "And there's the biodiversity issue. These trees are fast-growing, but they're not native. We could be creating monocultures that will eventually kill off local species. We're pushing nature, but it's not the kind of balanced restoration we need."

Another example is restoring peatlands are two examples of land use changes that can naturally remove carbon. However, this procedure is lengthy and would need vast amounts of precious land, which might forest an area the size of the United States, according to some estimates, and increase food costs fivefold. Additionally, the impact of carbon removal on trees is limited since unless they can be felled and burnt in a controlled system, they will ultimately die and release the carbon they have stored.

The Energy Debate

Raj stood up and started to present.

The potential of more affordable and effective nuclear energy has long been present with small modular reactors (SMRs). An industry that has long been beset by cost overruns and safety issues was supposed to enter a new age with its smaller, standardized designs.

But because no commercial SMR has yet been constructed in the US, skeptics are questioning the practicality of such technology as major tech companies like Google (GOOG) and Amazon (AMZN) look to them to fuel their AI goals with a minimal carbon footprint.

I don't want to forget to mention that," Raj paused for a moment" Since it may take them ten to fifteen years to construct, we are unsure of how successful they will be in combating climate change.

The global movement to shift away from fossil fuels in order to lessen the damaging emissions that are causing climate change has rekindled interest in nuclear power. Nuclear is still a desirable clean energy choice even if wind and solar electricity are widely available and reasonably priced. This is mostly due to the fact that nuclear can operate around the clock, year-round, and has a lesser environmental impact.

SMRs have shown the greatest potential. With a power capacity of 300 megawatts or fewer, modular reactors are one-third the size of conventional nuclear facilities, which have proved expensive and time-consuming. Because SMRs are manufactured in factories and put together on-site, the nuclear industry has praised their effectiveness and cost reductions.

There are just three SMRs in operation globally, two in China and one in Russia, despite the discussion of a streamlined procedure.

Aerial view of the construction site of Linglong-1 (ACP-100), the world's first onshore commercial small modular reactor (SMR), on July 4, 2024, in Changjiang Li Autonomous County, Hainan Province of China. (Wang Jian/VCG via Getty Images) · VCG via Getty Images

Here comes the challenges…..

Beyond the expense, there are further difficulties, such as the protracted regulatory licensing procedure and the issue of how to deal with all the radioactive waste.

SMRs would increase the amount of nuclear waste "by factors of 2 to 30," according to a Stanford University research, even if nuclear businesses will produce less trash if they keep a smaller footprint.

He looked around the table and continued, Some scientists are arguing that all these costs should be shifted to wind and solar energy. Others, however, retorted that nuclear reactors are more durable. Reactors can last 60 to 100 years, even if the initial expenses may be greater. Moreover, SMRs may be constructed nearer data centers because of their reduced footprint, which lowers infrastructure expenses.

According to the Department of Energy, nuclear energy is essential to the nation's move away from fossil fuels. The organization has allocated $900 million to support the creation of SMRs.

Although nuclear energy already supplies about half of the nation's carbon-free electricity, the Energy Department projects that the US will require an additional 700–900 GW of clean, stable power generating capacity to achieve net-zero emissions by 2050.

The group was silent yet again. The truth became undeniable: no matter how good a solution might be, there's a catch at the other end that might make things worse.

The Two-Edged Sword of Desalination

Meanwhile, water resources guru Dr. Ethan argued that the need for water is often exaggerated. "We've had to increase our desalination plants, especially on the coasts," he said. "They are taking seawater and converting it into freshwater more rapidly than we have seen it done in history and also along with a quick remedying of droughts."

Several heads nodded around the room. Clean water has become the new gold.

"But," Ethan continued, "The brine byproduct is another story. We're pumping it back into the oceans, and it's creating dead zones. The salt concentrations are killing marine life. We're trading one crisis for another."

The Underdogs' Awakening

The room settled into heavy, defeated quiet. Each scientist dropped forward, down to his or her notes, knowing that the solutions they'd fought so hard to develop weren't quite as life-widening brilliant as they'd once hoped. They had the knowledge, they had the technology, and they certainly had the drive in a collapsing world, none of it seemed enough.

Elena raised her head, observing the room with razor-sharp eyes. "We are doing everything we can," she said in a low, firm voice. "We knew from the start that we probably wouldn't make it. But we cannot quit now. Even though we cannot undo the harm which has been done, we are building a foundation for whoever follows us—however few they are."

"They'll have to learn from our mistakes," Raj said. "We've got to make a record of the technology, the science, and the shortcomings. If humanity survives, they'll need to know what worked and what didn't."

Mei nodded. "If they rebuild, they'll have to be smarter than we were."

David's voice was almost a whisper. "Maybe they'll listen next time."

For a moment, the scientists just sat there together in the half-light, united in their purpose, their burden. They were the final line of defense, the final wall between that world and utter destruction. And though they knew their work might never save the world, they would keep going. For the slim chance that something—anything—might make a difference.

Outside, the storm raged on.

Chapter VII

The Hidden Agenda

The sun hung low in the sky, casting a dim, orange haze over the city. Amina sat in front of her screen, the light from the monitor flickering across her face. Her heart was racing as she scrolled through the files that she had uncovered. For weeks, she had followed a trail of encrypted data, but now the pieces of the puzzle finally began to click into place. The truth was way worse than she could have imagined.

What… is this?" Amina whispered, inching closer to the screen. Her fingers hovered over the keyboard, too shocked to move.

It wasn't about negligence alone. Mistakes that government entities had made in handling the climate crisis were one thing. They were orchestrating it.

Between these records, Amina discovered a very unpleasant fact. The roots of today's disorder were placed then. Starting from the 50s, a number of powerful oil lobbyists were fully aware of the environmental impacts of fossil fuel usage. She felt astounded while reading a note written by the head of Air Pollution Foundation, the institution established by oil corporations back in 1953, which warned about the use of fossil fuels but was totally disregarded.

She started reading the article:

Large oil companies were aware of the harmful impacts of fossil fuels as early as the 1950s.
The chairman of the Air Pollution Foundation, which was established in 1953 by oil interests, issued a warning in recently discovered documents.

According to recently discovered records, major oil firms, including Shell and the forerunners of the energy behemoths Chevron, ExxonMobil, and BP, were warned about the impacts of fossil fuels on the planet's warming as early as 1954.

The Climate Investigations Center made the warning public, and the climate website DeSmog published it on Tuesday. The warning came from the leader of the Air Pollution Foundation, an organization founded by the industry. It may be the first time that big oil has been made aware of the potentially disastrous effects of their products.

Rebecca John, a researcher at the Climate Investigations Center who discovered the historic documents, stated, "Every time there's a push for climate action, [we see] fossil fuel companies downplay and deny the harms of burning fossil fuels." "Now, we have proof that they were acting in this manner back in the 1950s, when these very early efforts to combat pollution sources were underway."

In reaction to public concern about the fog engulfing Los Angeles County, oil interests established the Air Pollution Foundation in 1953.

Researchers had identified hydrocarbon pollution from fossil fuel sources such as cars and refineries as a primary culprit and Los Angeles officials had begun to propose pollution controls.

The Air Pollution Foundation, which was primarily funded by the lobbying organization Western States Petroleum Association, publicly claimed to want to help solve the smog crisis, but was set up in large part to counter efforts at regulation, the new memos indicate.

It's a commonly used tactic today, said Geoffrey Supran, an expert in climate disinformation at the University of Miami.

Supran stated, "The Air Pollution Foundation seems to be among the first and most blatant attempts by the oil industry to support a ... front group to distort scientific uncertainty in order to defend business as usual." "It contributed to the organizational and strategic foundation for big oil's decades-long delay and denial of climate change."

In the 1950s, the lobbying group, which was then known as the Western Oil and Gas Association, gave the Air Pollution Foundation $1.3 million, or $14 million in today's currency. This money was provided by member corporations such as Shell, as well as enterprises that were eventually acquired by or merged with ExxonMobil, BP, Chevron, Sunoco, and ConocoPhillips, as well as the utility SoCalGas in southern California.

📷 Oil refinery in Baton Rouge, Louisiana, owned by ExxonMobil, is the second largest in the country. Photograph: Barry Lewis/In Pictures/Getty Images

Lauren B. Hitchcock, a renowned chemical engineer, was chosen to be the president of the Air Pollution Foundation. And in 1954, the group, which up until that point had been claiming that backyard rubbish burning by homes was the cause, invited Caltech to submit a proposal to identify the primary source of smog.

According to a memo previously discovered by John, Caltech submitted its proposal in November 1954, containing important cautions regarding the coal, oil, and gas. It also stated that "a changing concentration of CO_2 in the atmosphere with reference to climate" might "ultimately prove of considerable significance to civilization." According to the recently discovered records, in March 1955, the Air Pollution Foundation informed the members of the Western Oil and Gas Association of the warning.

Climate scientists were starting to comprehend how fossil fuels cause global warming in the middle of the 1950s, and they were talking about their new findings in the press. The earliest documented warning to the oil sector about the greenhouse effect, however, was included in the recently discovered Air Pollution Foundation memo.

Funding for the Caltech project was granted by the board of trustees of the Air Pollution Foundation, which included members from SoCalGas and Union Oil, which Chevron ultimately purchased. Hitchcock, the head of the foundation, spoke before the California Senate in favor of state-funded pollution research after promoting pollution limits on oil refineries in the months that followed.

Leaders in the business chastised Hitchcock for his endeavors. The Western Oil and Gas Association informed him at a conference in April 1955 that he was pursuing "too broad a program" of study and bringing refinery pollution to the public's notice. According to the meeting minutes that John discovered, the Air Pollution Foundation was supposed to be "protective" of the business and produce "findings which would be accepted as unbiased."

According to papers examined by DeSmog, the foundation made no more mention of the possible climatic effect of fossil fuels after this conference.

According to Supran, "the fossil fuel industry is frequently perceived as having adopted the tobacco industry's strategy for suppressing science and obstructing regulation." "However, these documents indicate that, beginning with air pollution in the early to mid-1950s, big oil has been launching public relations campaigns to minimize the risks of its products for as long as big tobacco."

Many of the foundation's research initiatives were reorganized or redesigned to be carried out in close collaboration with lobbying organizations in the months that followed. In 1956, Hitchcock stepped down as president.

The Western States Petroleum Association was sued last year by Oregon's largest county for allegedly casting doubt on the climate problem despite having long been aware of it.

As previously discovered by DeSmog and the Climate Investigations Center, the Air Pollution Foundation funded the first studies on CO2 carried out by famed climate scientist Charles David Keeling in 1955 and 1956. These studies laid the groundwork for Keeling's seminal "Keeling Curve," which illustrates how fossil fuels raise atmospheric carbon dioxide levels.

📷 A gas flare from the Shell Chemical LP petroleum refinery burns against the sky in Louisiana. Photograph: Drew Angerer/Getty Images

According to other prior studies, large fossil fuel businesses investigated the effects of burning coal, oil, and gas on their own for decades. Exxon scientists had "breathtakingly" accurate forecasts of global warming in the 1970s and 1980s, yet they spent decades casting doubt on climate science, according to a 2023 research.

Caltech, the US National Archives, the University of California at San Diego, the State University of New York Buffalo, and 1950s Los Angeles newspapers are the sources of the recently discovered records.

Requests for response from the leading US fossil fuels lobby group, the American Petroleum Institute, and the Western States Petroleum Association were not answered.

All these calls remained unanswered, but the industry increased the expansion of oil and made more money at the expense of the planet. Amina then began to realize that the downfall of humanity was not destined to happen; it was however brought about by decision makers in skyscrapers who didn't care about what was taking place to the world around them.

"Ultimate death of us", she thought and continued reading…

Huge Oil and Gas Expansion Is Planned To Start In the Next Seven Years

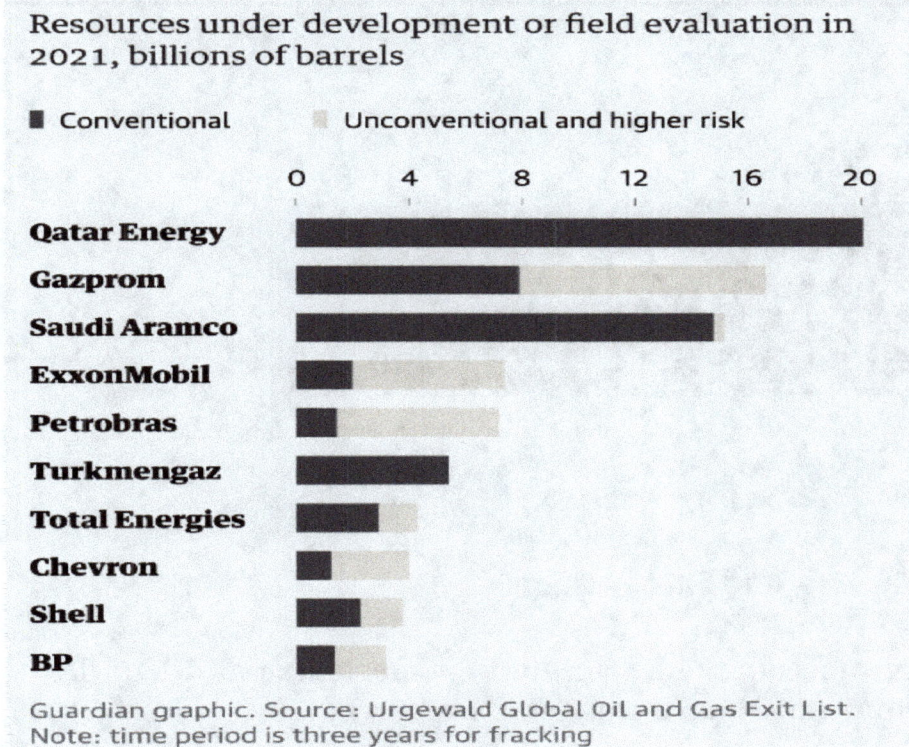

Resources under development or field evaluation in 2021, billions of barrels

■ Conventional   ▪ Unconventional and higher risk

Guardian graphic. Source: Urgewald Global Oil and Gas Exit List.
Note: time period is three years for fracking

The Middle East, US, and Russia Dominate Future Oil and Gas Production Plans

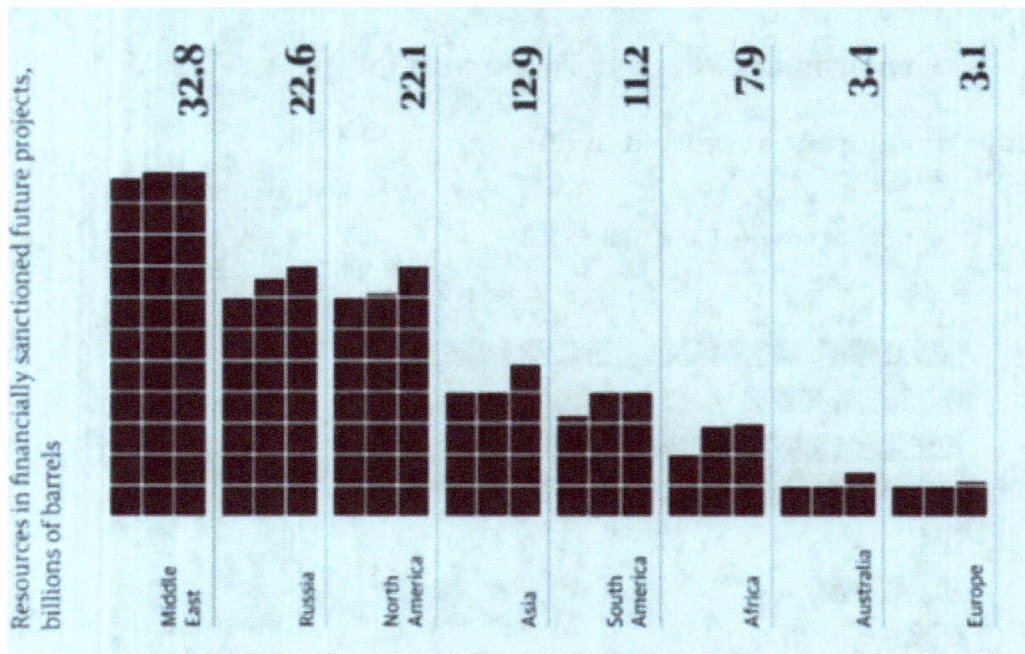

Resources in financially sanctioned future projects, billions of barrels

Middle East	Russia	North America	Asia	South America	Africa	Australia	Europe
32.8	22.6	22.1	12.9	11.2	7.9	3.4	3.1

Guardian graphic. Source: Rystad Energy UCube

Twenty-two mega-projects in the US account for more than a fifth of the Potential emissions from global carbon bomb

Potential carbon emissions from carbon bombs by country, GtCO2

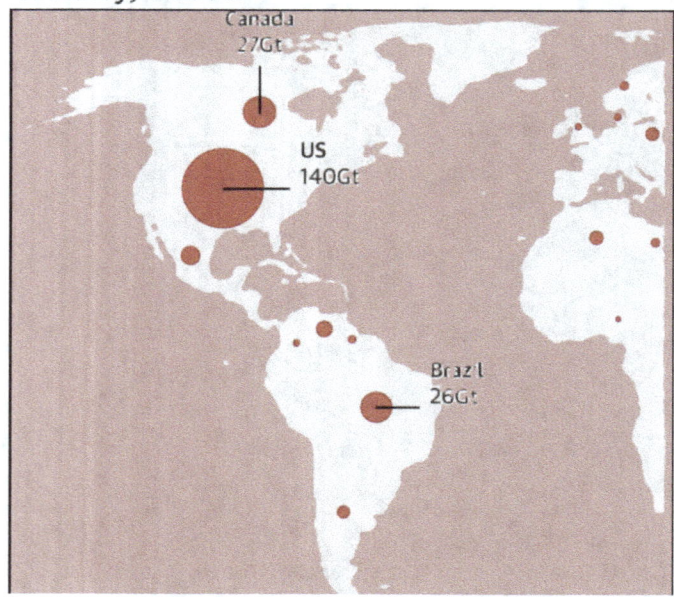

Canada 27Gt
US 140Gt
Brazil 26Gt

Guardian graphic. Source: Kühne, Energy Policy, 2022. MENA = Middle East and North Africa

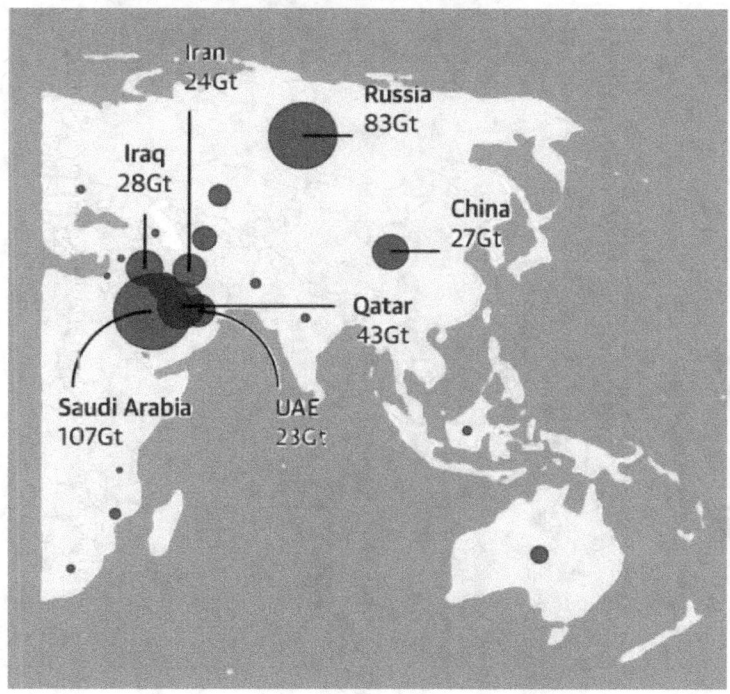

"Political leaders in wealthy nations can only have a colonial mindset if they decide that the billion-dollar profits made by fossil fuel companies outweigh the lives of the vast majority of black, brown, and impoverished people."

According to the assessment, the combined impact of these projects would result in 646 billion metric tons, or gigatonnes, of CO2 emissions, which would consume the entire global carbon budget. Over sixty percent of these programs are currently in operation.

The Leave it in the Ground Initiative's director, Kühne, stated that in order to prevent the world from collapsing even faster, the forty percent of projects that have not yet begun production must be halted. He also added that these projects ought to be a major focus of the global climate protest movement in the coming months and years.

"Despite the planet's burning, the oil and gas industry is still planning these massive projects." It doesn't seem like the lofty goals of the Paris Agreement were enough to cause them to reconsider their business strategy. The single biggest sign that we are not working hard enough is these carbon bombs.

Major companies plan to spend many millions a day to 2030 on exploiting new oil and gas

Capital expenditure per day 2021-2030, $m

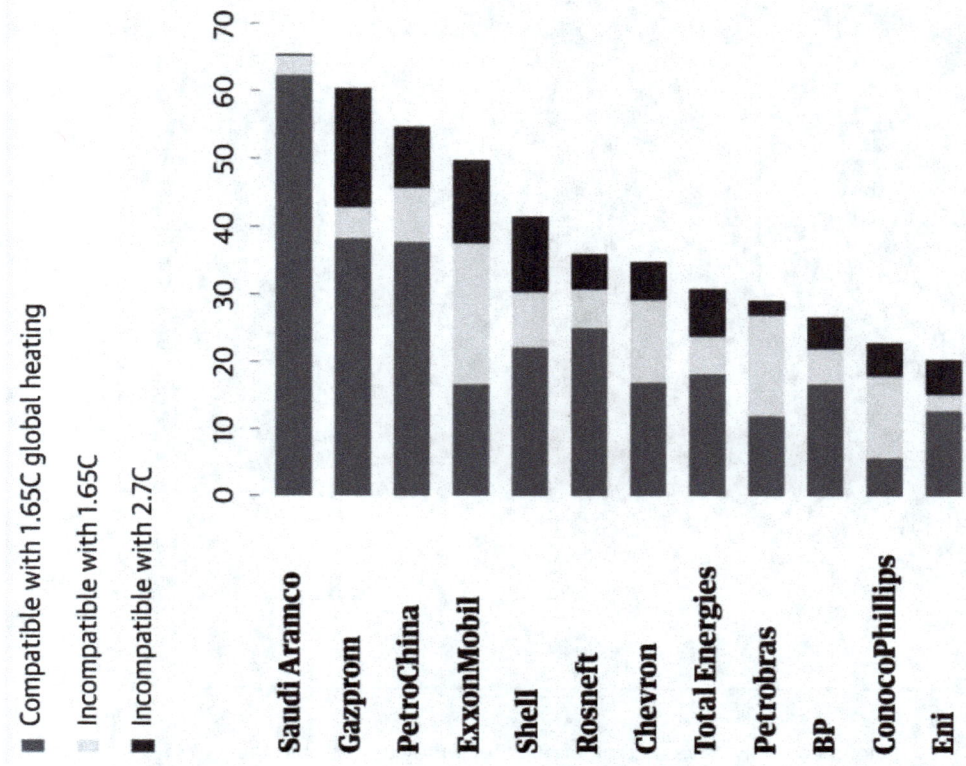

Guardian graphic. Source: Carbon Tracker Initiative / Rystad Energy

In the Paris Climate Agreement, international leaders committed to keeping global warming well below 2C and working toward a 1.5C increase in temperature. No new oil and gas developments are allowed in order to meet the latter's more stringent aim.

According to September's Carbon Tracker data, which represents the way below 2C objective at 1.65C, 27% of the corporations' expected investments are not consistent with this temperature.

With plans to commit $21 million a day until 2030, ExxonMobil has the greatest of these climate-busting investments, followed by Petrobras ($15 million), ConocoPhillips ($12 million) Chevron ($12 million) and Shell ($8 million).

Regarding the riskiest investments – those that might contribute to temperatures rising over 2.7C – Gazprom is responsible for $17 million daily, ExxonMobil for $12 million, Shell for $11 million and PetroChina for $9 million.

It was all there in chilling detail, the meetings between world leaders, chief executives of corporations, and defense contractors, discussions on how to manipulate climate data, not to safeguard the public, but to mislead them. The reports were fabricated, created to give the impression that the crisis was controllable, while it had been out of control many years prior.

Her pulse quickened when she clicked through another document, this one stamped *Top Secret*. It contained minutes from a private summit, the gathering no one was supposed to know about. The words on the screen made her stomach churn.

"If the public knows the truth, the panic will destabilize everything. Our economies and our governments will collapse. Controlling the narrative is our only option."

Amina's breath caught in her throat as she read further. They had been keeping technologies on purpose that could have slowed or even reversed the damage. Instead, breakthroughs had been buried and their developers silenced or discredited. Why? So that this elite could keep control of the world's resources, wealth, and power. Letting the world heal would have weakened that hold.

Then she saw the final, damning piece, plans for **safe zones**-private sanctuaries for the ultra-rich and powerful. Shelters built to withstand the very climate disasters the rest of the world would be left to face. While the planet was burning and the seas were rising, those privileged few would be protected, hoarding the last of the Earth's remaining resources when billions were left to die.

They knew. They had always known.

Amina's hands shook as she downloaded everything onto a secure drive. This was the smoking gun. The world needed to see this, and soon.

She grabbed her phone and dialed Ethan. "You need to come to my house … now," she whispered into the phone, her voice tight with tension. "I've found something huge.

Less than an hour, Ethan arrived, his face solemn as she handed him the tablet. As he read, the color bleached from his face.

"This is… this can't be real." His voice came out hoarse.

"It is," Amina replied, her tone even but icy. "The governments, the corporations-they're in it together. Lying to everyone. They've manipulated the data for years, and the climate crisis? That was allowed to happen on purpose. They had the solutions, Ethan. They could have stopped it, but they buried it because it would hurt their profits.

Ethan looked up at her, the weight of it all compressing on him. "The safe zones… they're going to leave the rest of us behind."

"They already have," Amina whispered. "The rich and powerful are readying themselves to ride this out while we burn. They've built entire sanctuaries for their protection, leaving the rest of us to fend for ourselves in a collapsing world.

Ethan sat down. His hands were trembling as he tried to wrap his mind around it. "This can't be happening. They can't just desert billions of people."

"They already have, Ethan. They've planned everything. The food shortages, the water crises, the storms-they're just letting it all take place. The world's falling apart, and they're watching it happen from their bunkers.

The room felt suffocating. Amina stood and began pacing, her mind racing with all she had read. The world was being lied to. The people trusted their leaders and believed the crisis was being handled, but it was all an illusion arranged with so much care.

People need to know," she said, her voice trembling with anger. "If we don't get this out, they're going to get away with it. We have got to make the world see what's happening."

Ethan shook his head, his voice was permeated with fear. "Amina, they've covered their tracks for years. If we leak this, they'll come after us. They'll kill us.

"I don't care," Amina spat. "The world is dying, Ethan. If we do nothing if we just sit by and watch, then what is the point of surviving?"

Ethan got to his feet, crossing the room to her side. "Alright," he said quietly. "We expose them. But we have to be smart about it. They have resources we can't even imagine. If they catch wind of this before we're ready…

"They won't," Amina said, cutting him off. "I've already begun copying the files. We'll send them to every underground news source, every whistleblower network. They can't silence us all."

The days that followed were tense. Amina and Ethan worked around the clock, gathering more evidence, securing encrypted communications, and preparing to release the files. But she couldn't shake the feeling that they were being watched. Every shadow felt like a threat, every stranger on the street a potential danger.

One evening, walking home, her worst suspicions were confirmed. A sleek, black car had been following her for several blocks. She quickened her pulse but could only force herself to walk calmly. Then when the car suddenly sped ahead and stopped on the corner, blocking her route, her stomach seemed to drop.

A tall man in a dark suit stepped out. His face was blank, unreadable.

"Dr Amina Hassan?" His voice was calm, almost pleasant.

"Yes?" she returned, trying to steady her voice.

"You're digging into things that don't concern you," the man said, taking one step closer. "It would be wise to stop."

Amina was frozen in place, her heart hammering hard in her chest. "I don't know what you're talking about."

The man smiled, but there was no warmth in it. "You know exactly what I'm talking about. We wouldn't want anything unfortunate to happen, would we?"

Without waiting for an answer, he turned and strode off back toward the car. Whizzing into the night, Amina's mind began to race. They knew. The elites—the very same orchestrating the climate collapse—knew exactly what she was doing.

That night Amina and Ethan were packing up their equipment, getting ready to hide. They knew the moment they dropped the files they'd become targets. But they did not have any other options. The world was running out of time.

As they packed Ethan paused, staring at a map spread out on the table. "Do you think it'll matter?" he asked softly.

Amina stopped, looking at him. "What do you mean?"

"Even if we unmask them… will it change anything?" His voice held doubt. "The safe zones, the control they have… they prepared for years in advance. When the world dies, they'll sit it out. And everyone else? Toast. What if it's too late?

Amina said nothing. Truth be told, she knew Ethan was right. The elitists had won a very long time ago. They'd built a world in which only the powerful few could exist. The rest? Well, those were just collateral damage.

But she couldn't stop. She couldn't give up. The world needed to know, even if it was too late and the world was going to fall apart.

She reached for the zipper on her bag and caught Ethan's gaze. "We fight," she said, "because if we don't, they will erase everything. They will erase us."

Ethan nodded, but the weight of what lay ahead was clear. The world was imploding, and they were standing at the edge of a precipice. The hidden agenda had put the planet on a path of destruction, and there was no going back now.

Walking into the night, they could not be rid of the feeling that the doom they were fighting to stop had already started.

Chapter VIII

Descent into Chaos

Amina prowled around the room, her mind pacing furiously. The weight of what she had discovered kept getting heavier and tighter as if it were a noose closing in on her neck. Ethan sat on the edge of the couch, staring up at her with apprehension, his foot tapping aggressively on the floor. Both of them knew they were on borrowed time.

We have to go public," Amina hissed, more to herself than to Ethan. "Unless we do, they'll just rub us out like they've done with anyone else who got too close before.

Public?" Ethan shook his head. In a low, urgent tone, he said, "You really think that's going to save us? They control everything--media, law enforcement, the web. We'll get discredited before we can even explain. Or worse, we just disappear. Just like that.

Amina stopped pacing and looked out the window. The night was bemusingly silent, an oppressive lull before the storm. She knew Ethan spoke the truth.

With every minute they wasted, those shadowy forces behind this conspiracy got that much closer.

Her hand instinctively reached for her bag, to the encrypted drives hidden there. Proof. The undeniable truth of how governments and corporations had been manipulating climate data for decades, suppressing breakthrough technologies that could have saved millions, and letting the planet spiral into chaos, all just to preserve their power.

The quiver was back in Amina's voice. "We are running out of time. If we don't act now, it won't matter. There won't be a world left to save."

Ethan stood up and crossed the room to her side. "And if we act? What if they stop us? What if—

A sudden crash of noise, and the door burst open, slamming against the adjacent wall as three men in black suits sprang inside. Amina's heart seized, and for a split second, she and Ethan locked eyes, terror flashing between them.

"Get down!" Ethan yelled, launching himself toward her, and thrusting her down to the floor.

Amina grabbed the drive from her pouch and ran towards the fire escape as her heart was racing. Ethan stayed behind to try to slow them down. His voice was quickly overpowered by the sound of gunfire.

The cold night air stung her face as she started to climb down the metal ladder, every movement faster, more desperate than the last. Behind her, she heard Ethan yell in pain—a sound tearing through her soul. But she couldn't stop. She couldn't look back. Ethan had bought her time, and she had to make sure it wasn't in vain.

When she reached the ground, her mind was racing on pure adrenaline. She knew it would only be a question of minutes before they caught up with her. The men chasing her were not just hired muscle, they were professional men who taught in the ways of eliminating anybody who stood in the way of the agenda.

She slipped into the alley's shadows, her chest heaving as she tried to catch her breath. She had nowhere to go. Nowhere was safe. Ethan was gone, and the last shred of hope she had clenched her fist tightly. Unconsciously to save the drive.

The world around her weighed her down and didn't let her breathe. The streets were eerily silent, as if the city, too, knew the storm would come. The storm nobody would be able to stop.

It took Amina hours, but finally, she reached the safe house Ethan prepared for them many weeks in advance. The lowly apartment above the slums was not a place where either the rich or powerful would look for her. She fell onto the floor, having locked the door behind her, exhausted and shaking.

She slotted the drive into the secure terminal Ethan had rigged. Her fingers flew across it, uploading the data to every possible source she could think underground news networks, whistleblower sites, and forums that still operated in the dark corners of the web.

As the documents downloaded, Amina allowed herself a momentary flash of optimism. Maybe this would be enough. Perhaps the truth would spread faster than they could suppress it. Maybe, just maybe, the world would wake up in time.

But as she watched the files go up, gnawing doubt crept into her mind. It was too late, wasn't it? Even supposing the world saw what she had dug up, would it make a difference? The elites had prepared for this. Safe zones, hidden stockpiles of resources, and secret technologies to shield themselves from the coming chaos.

The rest of the world? They would be left to burn. Days passed and Amina stayed in hiding, waiting for any sign that the penny had finally dropped. But instead of public outcry, instead of revolution, the news channels maintained their silence. The governments kept feeding their people the same meticulously concocted lies. The climate crisis got worse, it seemed the masses were still too lethargic to rise to the cause.

Then the riots began.

They started as small, isolated pockets of unrest-demanding people wanting answers, desperate for solutions. But the truth Amina had uncovered never reached them. The protests continued to grow, spiraling completely out of control. In response, governments declared martial law. In the streets, soldiers lined the sidewalks, imposing their curfews by lethal force.

With growing unrest at their doorstep, the elites accelerated their plans. Cities were locked down; borders sealed. Only the wealthy and powerful could flee into the hidden safe zones, leaving billions of others to face the growing storms, the rising seas, and the choking heat.

Amina watched, through her window, as the world fell apart.

Then, the skies turned grey, and the once bustling streets beneath the skyscrapers turned into battlefields. Buildings went up in flames, and people snarled over leftover bits of food while the air was thick with desperation. Amina could see it now, collapse wasn't coming. It was already here.

The day finally came when they found her. She had known it was inevitable. She had been watching as the world slipped further into chaos, powerless to stop it.

The door is knocked. Cool and measured. They didn't have to break this one down.
Amina did not attempt to conceal herself. She sat at her desk, the room was deep in shadow. What she could do, she had done. And it hadn't been enough.
The door opened. Two men came through. The same two men who took Ethan.

"It's over," one of them said, his voice was cold, almost indifferent.

Amina looked up at them, her eyes hollow, and her hope long gone. "It was always over," she whispered.

The man stepped closer and reached for the drive still plugged into the terminal. "You should have stayed quiet," he said, pulling it free. "The world is beyond saving now. You just couldn't accept that."

She closed her eyes as they approached, crushed by the weight of her failure. The hidden agenda had worked. The powerful had secured their future while the rest of humanity was left to fend for themselves in a dying world.

There was nothing more she could do to stop it.

The world burned, and Amina was just another forgotten casualty among the ashes.

Chapter IX

Ashes of the Fallen World

For as long as we can remember, time has always been strange. Slipping through our fingers when life is good, flying by unnoticed during moments of happiness and contentment. It felt like an eternity when the world went dark and survival became a daily struggle. We felt as though time lingering had amplified our suffering as each second felt like a slow crawl through agony. While we desperately wanted it to end, it was as if time itself had turned cruel.

No matter if it's flying or crawling…. It is a slow march towards the inevitable demise.

Michael went through the same experience. He lost track of the days, weeks, and months.

Michael's hands were shaking as they reached into the trash can, overflowing with garbage. His once-powerful, calloused fingers were brittle and grimy, traversing the mire as if they belonged there. He thrust aside rotting food and plastic wrappers, the stench rising but, he hardly noticed it anymore. Hunger had deprived him of the luxury of disgust. There lay a half-eaten piece of bread, green with mold, he hesitated for only a second before pocketing it. He had learned not to be choosey.

The mold was another reminder of just how far down the chute he'd slid. Once, Michael had been that man of success—a husband, a father, and a man who never had a care about where his next meal was coming from. Their house had been a beautiful fortress of comfort, full of laughter and life. But the world had just spun and unraveled into chaos before anyone knew what was happening.

All of that was gone now.

His stomach had twisted with pain as he recalled that day they'd tried to leave the city. It felt like a lifetime ago, but the memory was still sharp, like a wound which refused to heal.

The collapse began with whispers on the news-at first, vague warnings of climate disasters, power shortages, rising food prices, and political unrest.
But soon, the whispers grew deafening. Michael recalled the day it struck him like a punch to the gut. All channels were loaded with the same macabre headlines, all chanting the same bleak statistic, **"Science tells us that within 100 days of total collapse, 75 percent of humanity will die."** The voices of news anchors can still be heard shaking as they read the words, their eyes hazed by the same fears playing inside everybody's head. Satellite images flooded the screens with burning cities throughout the world, riots, starving hordes, decaying bodies, burning houses, and flooded homes. Clips of scientists, pale and hollow-eyed, confirmed the worst.

Global supply chains had snapped like brittle twigs, the disease was spreading through crowded camps, and the earth was warming faster than anyone had predicted. Michael still heard the reverberation of that figure "75%" as that landed heavily in his chest like lead. That was when the world stopped pretending that things would get better.

To begin with, Michael did not believe the news.

The apocalyptic report of global fall down, riots, and mass starvation felt too surreal, too far from his comfortable life to begin with. He would scan through the channels until something glimmered, a flicker of hope showed up, but all the broadcasts came as versions of the same disastrous news. The doom-filled prophecy broadcasted over and over again as if reality had cracked out overnight. Frustrated, Michael finally snapped it off and decided to take things into his own hands. "The media always exaggerates", he thought.

There must be somewhere safer. He began dialing his old friends and clients all around the globe, hoping that at least one of them could help get his family out of the nightmare that was closing in on them. Only to ring his business partner in Dubai, Ali, whose voice trembled with fear as he described riots breaking out in the streets and food running out. Michael tried Europe, dialing an old friend in London, but the answer was the same, chaos, violence, and bare shelves in store. Not even in China, where he had conducted business with rich contacts, was there any respite.

His friends talked about falling infrastructure, crops dying under the intense heat, and general panic. Every call ended this way, the world was falling apart, and no one, regardless of where he lived, was safe in the collapse. Michael hung up the phone each time he finished talking, the sinking realization slowly crossed his mind - "nowhere left to run".

They were all trapped in the same nightmare, and the walls were closing in. In the early days, when the world's unraveling still felt like something that could be stopped, Michael clung to the hope that governments would intervene.

After all, surely the people in power—those elected to protect their citizens—would act before things went too far. He believed in the system, in the promise of leadership, and thought somehow reason would prevail. But as the weeks went by, the realization dawned that no rescue was coming. Michael's hope gave way to anger, and he started to dig deeper to find out why no one was doing anything to stop the chaos. He had always been a man who trusted facts, so he turned to research.

The more he read, the more disillusioned he became. In the West, he discovered, democratic governments were shackled by the influence of wealthy donors and powerful corporations. Politicians—no matter how well-intentioned they may appear—were too beholden to the industries that funded their campaigns. Big businesses—fossil fuel companies, pharmaceutical giants, and tech firms—dictated policy from behind the scenes.

Environmental concerns, social justice, and long-term survival were little more than afterthoughts, on the altar of short-term profits. Wealth creation became the real enemy, propelling decisions that kept the world firmly on a collision course with disaster. He saw it now, clearly. The leaders weren't acting because they couldn't. They were trapped, unable to challenge the industries that kept them in power. They were afraid to rock the boat, for they feared the world itself might cave in.

But what his eyes chanced upon the eastern shore filled him with sorrow.

He found more dejection there. Governments in the Middle East, with oil riches bountiful, got themselves trapped in their own concoction. Their economies were dependent solely on the fossil fuel industry, the same fossil fuel industry strangling the very life from the planet. These nations, formed on the promise of a bottomless pit of oil money, could not afford to pivot away from the resources that brought them their power and influence.

Like their Western counterparts, their leaders played a futile game without an end—in defending an industry making the Earth uninhabitable day by day. The simple fact is that nature does not compromise. Global warming is not something to be addressed in negotiations between politicians or economists. It is something of a force, once harnessed past a certain point, that cannot be turned off, will simply gain steam and make the planet unlivable.

And for the first time, Michael realized no one was coming to save them.

After all, He was a human, how he could go against the deep psychology of a human being?

He wanted to survive…….**At any cost….**

He thought maybe if he left the city and retreated to the nearby forests, he could wait it out—just for a few days or weeks—until everything calmed down and life returned to normal.

Michael had packed the car with essentials to survive. Supplies, food, water, and blankets. His pregnant wife Amelia clutched their daughters tightly, trying to be calm for them. Little did the family know that even Lena, Alice, and tiny Emma could feel the fear starting to spill from their parents' pores, though no one had ever.. ever spoken the dreaded words aloud.

'We'll be fine when we leave the city," Michael had said. His voice was steady, but inside he was coming apart at the seams. "It will be alright. We just have to get out of here".

But it wasn't okay. The roads had been chockfull of thousands of cars all trying to escape. People were abandoning their vehicles, the highways turned into graveyards of empty gas tanks and broken dreams. Their car had spluttered to a halt in the middle of that mayhem, and Michael felt something inside of him snap when he saw the flashing red light on the gas gauge.

They had nothing left to do but join the crowd of people that walked through the city, hearing the sounds of collapsing civilization.

Looters by then had pillaged the stores, and gangs walk up and down like predators hunting whatever is left to scavenge. It was now evident that nobody was coming to their rescue. The government appeared frozen, much like the people. Politicians, bound to their donors, with their actions ruled by the bottom line, had sat at their desk and watched the world burn. Now, action of any kind was too late.

The same thieves who had looted the stores had moved into the neighborhoods, robbing homes, stealing everything of value, and burning the rest. Michael would never forget seeing his house in flames in the distance. Amelia could not differentiate whether she was feeling sad, angry, helpless, or nothing. All her memories and dreams were reduced to smokes". Amelia cried silently that night, holding their daughters close as they slept on the cold pavement. He wanted to comfort her, but what could he say? There was no comfort left to give.

Days blurred together as they walked, moving from one street to the next, where more horrors revealed themselves around every corner.

They were dragging people out of their houses by arm and leg to share some crumbs of food, and killing over water. No more rules. No more laws. The police had already run away; the government has crumpled under its own impotence. Every man, woman, and child was standing alone.... alone in the fight for survival in a world that no longer cared. In the end, they found the refugee camp-a sprawling wasteland of tents and shanties.

It was nothing like what they had imagined, more like a prison without walls, with desperate people clawing at each other to survive. Violence erupted over the most basic necessities in air heavy with human waste. People fought for a sip of water, a place to sleep, an opportunity to use one of the few crumbling toilets. It was daily struggle for a right to exist.

He had desperately tried to keep his family safe. Miles and miles of walking each day only for food and water were what kept Amelia and the girls alive. Lines he never thought possible crossed. Like so many in the camp or out of it, he tried to help his family.

One day, he stole a stale piece of bread from a starving child in the camp, the next, he beat an old man to death—for what?
A few rotten bananas. He did it all for Amelia, hoping that those scraps might give her enough strength to hold on a little longer and survive until things returned to normal. Michael reached the second stage of human desperation where survival outweighed morality and acts previously held to be unthinkable were done. Dignity and decency could no longer be afforded.

Yet it was never enough. The meager supplies he found were deplorable, deteriorating leftovers, rotten vegetables, or water with mud suspended within, barely good enough for drinking. Amelia's hunger-weakened body had become increasingly ill each day. The little baby inside her had totally drained what little strength she now possessed, and there was no one, no physicians, no medicines to treat her with.

Michael had held her hand as she breathed out her last breath. She mumbled his name, only a little louder than a whisper, her eyes filled with fear for their daughters. And then she was gone. She died silently like Many hundred thousands of children and weak men and women around the camp or it would be more precise to say around the world......Her death had been silent, the way she'd fallen to the ground from his grasp. But the anguish that followed, deafening. He buried her in dirt outside the camp, alone.

And the girls—Lena, Alice, Emma—they had been so silent after that. Too silent. Michael knew they were scared. He was afraid himself, but grief was not allowed. Survival was the only thing that counted now.

Then the raiders came.

It was just another hot day, and the air was unbearable, heavy with tension. Armed men stormed through the camp, faces cruel and frosty, seeking whatever of worth there was. But it was not food or drink they wanted. Nor was it money—they'd find little use for money in a world that was crumbling. No, they sought something else. Something far worse.

They took his daughters.

No! Please!" Michael had screamed, his voice cracking with desperation. He'd flung himself upon the men, pleading with them to take him instead. He'd offered them everything he had left. Money, gold, even the secret stash of wealth he'd buried before they'd abandoned their home. The men merely laughed. There was no longer any use for money. All that mattered now was power. And they had it all.

They took his daughters. Michael tried to fight everything within him. He screamed, clawed, begged. Nothing helped. They left him in the dust, bruised and broken. His body hurt inside his mind breaking. When he opened his eyes, they were gone.
Time stopped existing after that….. He looked for them, questioning every passerby, every survivor he encountered, but none had ever seen them. The daughters had fallen into the same abyss that had swallowed the entire world. He walked around, invisible and skeletal, through the ruins of a dying civilization, as his body withered and his mind unraveled.

But something darker was growing inside him, day by day. The world had become ash, and Michael had nothing to lose anymore. He found other people-men like him, broken and angry, men who had lost everything and were willing to fight. Together, they became something more dangerous, something the world had never seen before.

A new plan began to formulate in his head. He wanted only one thing at first-that is, to find the girls, holding onto the desperate hope that there was a place out there where they still breathed life.

As days became weeks, so did he learn a little more about the men who were accompanying him.

Among them were the ones who used to work for the rich, the ones who built the underground bunkers designed to shield the already-exclusive elites from the hell breaking lose at the surface. They knew exactly where these secret fortresses were stocked with food, water, and luxuries devised to shield the elites from the collapse. But what the rich hadn't anticipated was that those who were to build their bunkers would turn on them. Michael decided that if he could not find his daughters, he would at least make those who thought they would be able to escape this disaster suffer like everyone else. They would strip open these underground vaults, tear apart their safe havens, and take what they needed to survive.

These bunkers, symbols of human prosperity and wealth's apogee, would be nothing but tombs. The poor, entombed underground, would be no less exposed to the ruin than those who lay abandoned above in the open air. They would just die slower, in the dark. They attacked the high-ranking bunkers, razing in what remained of the last survivors in a world that had left them for dead. The rich, who thought to ride out the storm in their bunkers below, were no more secure than any soul. Michael's lot busted open the bunkers like tin cans, ransacking everything they had stockpiled-food, water, guns-and burned up what was left. It was nauseating irony.

They thought they'd bought their way to immunity from the apocalypse. They were just as dead on arrival as everyone else. And gold, money, power - it was all pittance now. The new currency in this world was life, and life involved doing whatever you needed to take it, no matter how repugnant.

As Michael stood amidst the flames, observing the burning of another bunker, he didn't feel anything. Hollowed by the rage he could not hold in his chest, nothing mattered anymore. He had fought so hard, it had all been for nothing. He had not found his daughters. He hadn't brought Amelia back. All he had was the fight.
In the end, even that wasn't enough.

The world was dying, and everybody in it was doomed.

The governments, the corporations, the industries—they had all fought for their interests, but nothing mattered anymore. As the planet burned and temperatures rocketed higher than any human ever survived, the earth itself appears to be swallowing all of them in. The pursuit of profits brought about their own downfall, and now it is too late to change back. Michael knew, in his heart, that nobody was going to make it through this. Not for very long. The fat cats would hole up in their air-conditioned bunkers, and the rest of them would fight it out on the streets, but when the heat spiked to 125 degrees, and the crops started dying, and the water stopped running— everybody would die. Some would go down into the earth, living like cavemen, hanging onto life down there in the dirt, but what kind of life would that be? Without resources…. huh.

In those early days, Michael was part of that crowd who believed in the system. He thought governments could balance the needs of their people. But as time went by, he began to notice cracks/subtle at first, then widening into crevices of corruption and self-interest in people, businesses, and institutions carving out their slice of the pie for personal gain over collective good.

This was almost silent destruction, a slow and steady unraveling. The mighty few clamored for laws that served only the interests, businesses lobbied for benefits that assured them shelter from competition, and institutions clamored for special privileges. These events were not isolated, they set a pattern for every corner of society. The government, caught up by these competing demands, lost the capacity to act in the best interest of all.

Michael remembered what his father used to say "The needs of the many must outweigh the needs of the few." But in the final years, those words had been forgotten. Every decision aided the elite while the majority had to scrounge and fight for whatever they could get. As the mighty grew more powerful the government grew weaker its purpose was buried under the weight of individual ambitions.

Global warming had been the tipping point. Scientists had been threatening to destroy it with devastating consequences if left unchecked for years. Those in power had delayed, not willing to make adjustments that would bear a dent in their bottom line. When finally taken into action, it was too little, too late. The cities drowned, crops withered, and chaos spread. The very system that was to protect the people had doomed them instead.

Michael clutched his fists together, the anger welling inside. He has seen it all, the erosion in collective responsibility, the government paralyzed by the demands of a few, and the collapse that followed. Ashes beneath his feet were what buildings were made of, but it was also what a society was made of when it forgets its survival depended on many and not a few.

Now, standing in the ruins, Michael knew it was not the story of one government, or any one particular time but rather the silly nature of all societies who moved priorities from the greater good to self. And as long as that lesson didn't learn itself, history would repeat and leave nothing but ashes.

Chapter X

The New Human

The world was at its breaking point. A hundred days had elapsed since the final collapse, and Kade roamed across the deserts which once were the bustling cities. The skyscrapers had all but withered to bone fragments, their broken forms a testament to the fragility of human ambition. Only a few remained, living with terror in their hearts, before a burning sun. The temperature during that time was unmeasurably high making it possible to survive only underground in chasms. Life as they knew existed nowhere above the ground.

Kade had therefore become accustomed to the dark and almost hostile environment of the hidden bunker under the earth; despite these circumstances, he often got bored and attempted outlandish and risky journeys on the surface of the Earth. Each time he moved out of the cabin, like a man coming out of a convection oven – 125 degrees Fahrenheit, dry, without air, and without mercy. He would be camouflaged and withdraw to any available dark area as he went about the destroyed civilization. These Bob Cratchets began carrying as much water as they could and drank sparingly, but the heat had always prevailed. His skin became red hot, he couldn't see very well, and realizing he was parched, he drank as much water as he could, but he got heat stroke nonetheless. Each time Kade went to the surface of the bunker his body fell to the ground covered in sweat and it would take him days, sometimes weeks to get well from the beating that the surface had given him. However, the situation gave him no incentives to stay away; something in him compelled him to jump back into the ruins, to seek solutions in the ashes of a conquered reality.

Caves and bunkers that saved them seemed to offer little comfort. They were prison-like havens with little or no light, fresh air, or the extravagance once assumed to be limitless. Centuries of medicine, technology, and conveniences had disappeared. Life had become something primal and raw. Kade's days were full of survival-base cravings-food, the preservation of water, and trying to avoid getting ill in a world with no medicines.

Lighting was an essential lifeline in the bunkers while at the same time being a constant life-threatening factor. Most bunkers were powered by old generators, using fuel that they had acquired earlier before the collapse; diesel, natural gas, and sometimes solar power harvested from solar panels installed on the surface. However, fuel was scarce, and the generators were collapse-prone – quite unreliable. The survivors lay awake day and night thinking of what would happen when the power supply was to finally switch off. They could not afford to use electricity often only to light their rooms, ventilate the house, and for the greenhouse where they grew their food. There was always fixing something, stripping the carcasses of technology and countless times they only had almost no light, a constant reminder of how close they were to being dead.

This meant that every time someone got sick it was bad news. They had no doctors or access to modern medicine; they only had some hoarded medical stuff—the old antibiotics, painkillers, and the barest forms of first aid kits. Such supplies were exhausted very fast and later, sickness was equivalent to death. To the onset of minor indications such as cuts, bruises, and most of the common ailments, people attempted to Self-treat but for other severe complications, there was no expectation of recovery. Fever, infection, or disease could spread through a body, and all that one could do was wait untreated by his side.

Childbirth was especially deadly. When a woman required surgical intervention to give birth, there were no clean rooms, no tools, and no surgeon who could assist. There was no personal protective equipment; instruments were available; the procedures had to be done in barely lit, unsterile conditions; and equipment was sterilized over fires. It was brutal and most times deadly. Childbearing was a huge risk for many women, and the few fortunate to give birth had little assistance in regaining their strength.

Likewise, newborns' lives were projected to be almost unlivable as well. Due to inadequate access to medical attention, many babies would even die. First of all, there was too little warmth, nutrition, and medicine for premature infants. There were no incubators, no vaccines, and no bottles with the formula. Many babies could not even breathe or swallow well. Their parents could only watch them fade away. Those who were lucky to survive were mostly undernourished, growing up in a world with a high risk of getting sick and dying. The bunkers which were created to shelter these people and save them, have turned into dungeons of gloom, where existence was as weak as it could be.

But even in this desolate quiet, something extraordinary was happening. Nature, long oppressed by the hand of man, was beginning to reclaim its territory in really unexpected ways. Even with unbearable heat, some resilient species had started adapting themselves. Insects, reptiles, and small burrowing animals, so long in tune with adverse conditions of the environment, were finding a way to survive. They lay low at night or dug deep in the dirt to avoid the lethal heat of the day. Plants also began to sprout, those droughts and extremely hardened things, splitting the concrete cracks, and seeping through the rubble.

Changes of scenery caught the attention of Kade as well. He watched how the recently dead land was now splattered with heat-resistant grasses and desert flora. They did not require much water but survived without human cultivation. They have been reclaiming that which was lost. The urban decay, which once smothered the rest of the earth, was being pushed back.

Few animals were roaming the above ground, but they were hardy. At night, when temperatures merely dipped low enough, Kade saw all sorts of creatures come out. There were birds he had never seen before flying across the sky to where their nature had traditionally required them to be. Small mammals scurried about, adapted as they were to the conditions. It seemed that life, even without humans, found a way to adapt.

Subterraneous, too, the natural world continued to thrive. Microbial life, fungal life, and insects teeming multiplied, breaking down fragments of human civilization in order to rebalance matters. They formed the bedrock of a new ecosystem- a slow, powerful force working to heal the Earth at ground level.

He finally understood that the Earth, with all its inhabitants barely surviving, was healing itself. The Earth had gone through extreme climates before and would go through this one, too. It would take hundreds or even thousands of years for the Earth to rebuild, but with no human presence.

As Kade wandered through this strange new world, he sometimes would ask himself: was man a cancer on Earth? It troubled him as he saw the natural world rebound from the wounds inflicted upon it by humankind. They had used and ravaged the planet to extinction for hundreds of years. And yet, seemingly on its own, it seemed healthier without man as a part of its skin. The few remnants of surviving human legions consisted of few, scattered, and savage versions of what once was the dominant civilization on Earth.

Kade thought back to the stories his mother had told him. Economics, she had said, was designed to improve lives. But in the end, it had become a machine of destruction. It had been doomed from the very beginning to fail because of infinite systems on a finite planet. His race's idea of endless growth, where the strong take all, had brought about its downfall. The Earth had revolted, and the systems they set up to make life better were set in motion for their complete destruction.

Through the waste, Kade thought of the bees and ants that his mother once mentioned. They survive on cooperation, and they sacrifice for the greater good. Humanity was driven by selfishness and greed. It was an experiment in intelligent evolution that had gone terribly wrong and came at an unbearable cost. Such a simple question wracked Kade's brain: did humanity deserve this fate? Was collapse an inevitability brought about by their inability to adapt to the realities of a finite planet? Or perhaps the collapse, in itself, was Earth's way of showing humanity the exit door-a final judgment on all of that human exploitation and over-exploitation over the centuries?. Or perhaps, just maybe, there was something worth saving in the existence of humanity. Kade wasn't certain. Could there possibly evolve a new and superior species of human being, one which would learn from the mistakes of the past and find a proper sense of balance, of communal sacrifice, and of harmony with the Earth? It was a question without an easy answer, but it was one that Kade knew he had to wrestle with.

Kade stood among the ruins, contemplating the universe: infinite, indifferent, and vast. Would it care if humanity blinked out of existence? Would the stars pause for even a second if the Earth became another Venus, a lifeless furnace doomed to burn forever? It was rhetorical, a question that opened the door to endless debate. And yet, in his bones, Kade felt that there was something small though it was worth saving.

Perhaps there was hope for humanity to heal also. It would just take a new kind of thinking, a new kind of human-one who learned from the past, embraced cooperation over competition, and understood his place in the world wasn't to dominate, but to coexist. If such a human could rise from the ashes, perhaps there was still hope.

Chapter XI

A New Frontier

Around the world, communities once thriving were becoming impossible to live in due to extreme weather events that were getting worse. For most, soaring heat and historic humidity had become unbearable. There were tales of regions in South Asia, the Middle East, and parts of the southern United States where heatwaves, with humidity at nearly 100%, had become deadly. Those "wet-bulb" conditions make it impossible for the human body to cool itself through sweating. In such places, survival became a matter of minutes, not hours; and the survivors realized they had fled in time.

Rahim sat silently tormented by memories of blurring heat waves back in India and excruciating Siberian chills. He would recall the feeling of being a prisoner in a sauna on those sweaty, sultry days with the heat and humidity climbing endlessly. The situation was just like he was sealed in a room when, in fact, he was trying to open a door that existed only in his imagination. But here, now in the frozen wilderness, nature had merely swapped one brutal extreme for another. "Is there anywhere left on Earth that isn't trying to kill us?" he thought, as he watched his father struggle against the cold.

He stared at the world map he once pasted on his room wall, marked with the most humid cities on Earth.

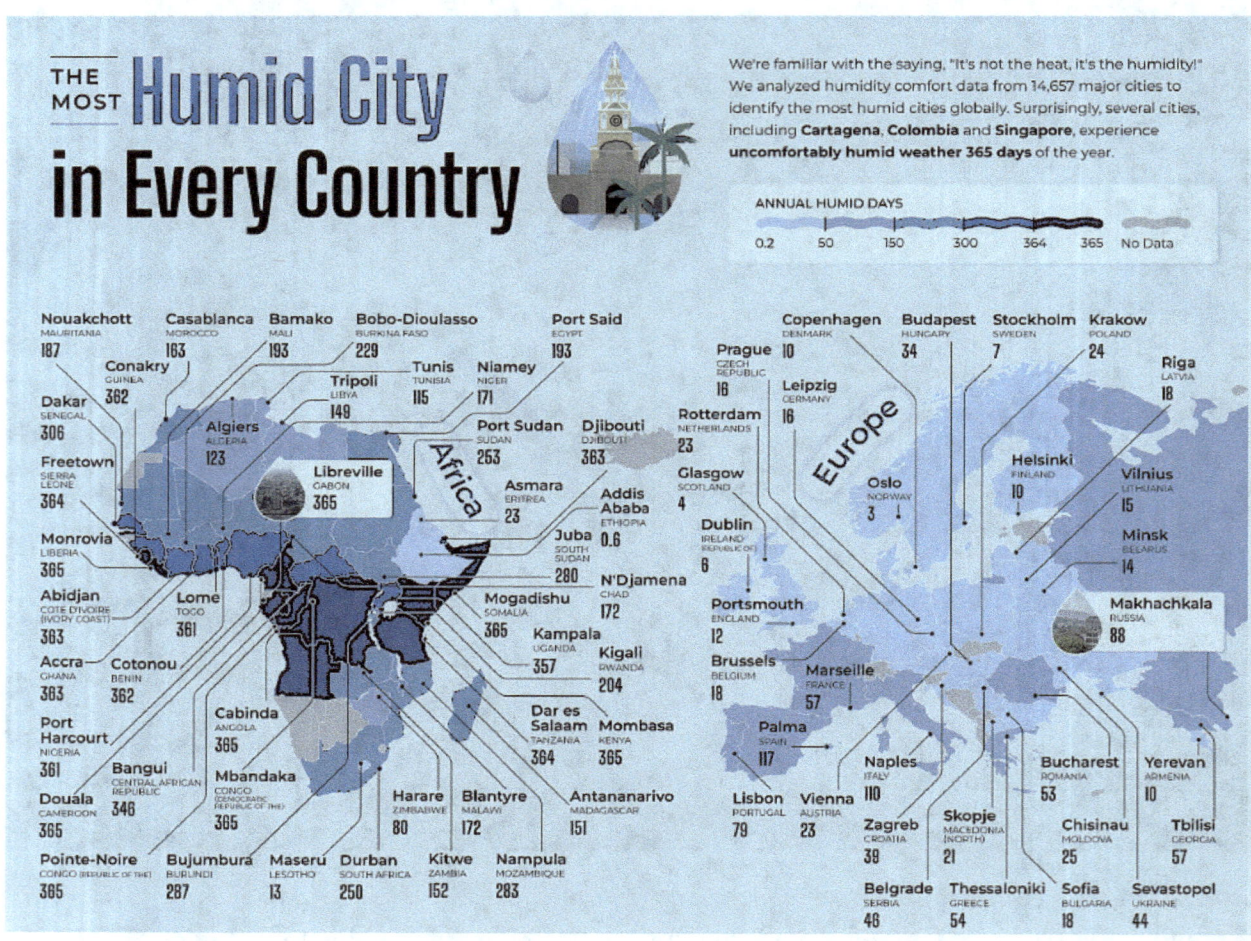

Methodology: We used dew point temperature data from **WeatherSpark.com** to rank 14,657 cities worldwide based on the number of uncomfortably humid days in 2023 with a dew point of 60°F (16°C) or above. Ties were resolved by considering extreme humidity days (75°F/24°C dew point or above).

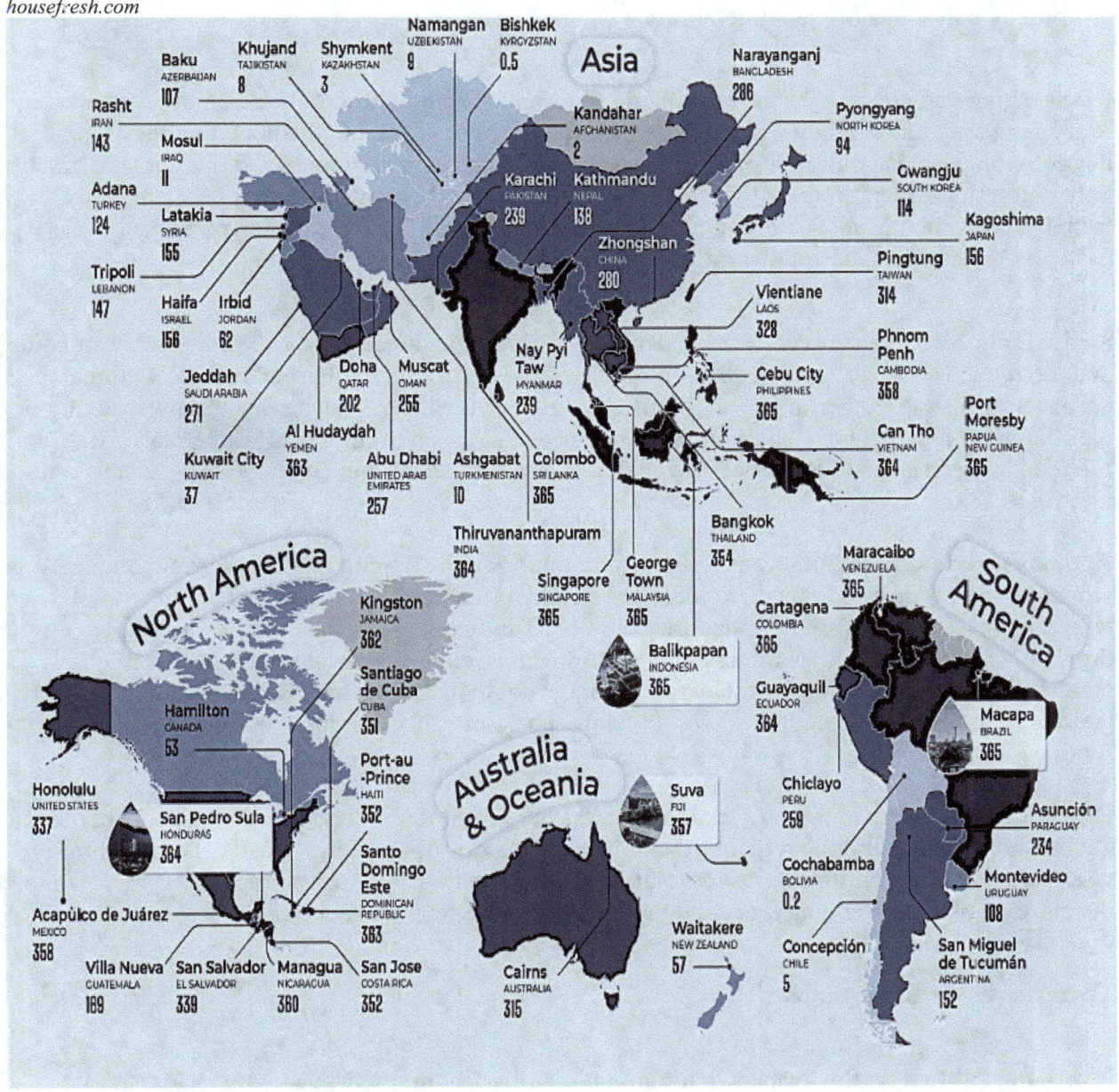

Asia

Namangan UZBEKISTAN 9
Bishkek KYRGYZSTAN 0.5
Khujand TAJIKISTAN 8
Shymkent KAZAKHSTAN 3
Baku AZERBAIJAN 107
Rasht IRAN 143
Mosul IRAQ 11
Adana TURKEY 124
Latakia SYRIA 155
Tripoli LEBANON 147
Haifa ISRAEL 156
Irbid JORDAN 62
Kandahar AFGHANISTAN 2
Karachi PAKISTAN 239
Kathmandu NEPAL 138
Narayanganj BANGLADESH 286
Pyongyang NORTH KOREA 94
Gwangju SOUTH KOREA 114
Kagoshima JAPAN 156
Zhongshan CHINA 280
Pingtung TAIWAN 314
Vientiane LAOS 328
Phnom Penh CAMBODIA 358
Cebu City PHILIPPINES 365
Can Tho VIETNAM 364
Port Moresby PAPUA NEW GUINEA 365
Jeddah SAUDI ARABIA 271
Doha QATAR 202
Muscat OMAN 255
Nay Pyi Taw MYANMAR 239
Al Hudaydah YEMEN 363
Kuwait City KUWAIT 37
Abu Dhabi UNITED ARAB EMIRATES 257
Ashgabat TURKMENISTAN 10
Colombo SRI LANKA 365
Bangkok THAILAND 354
Thiruvananthapuram INDIA 364
Singapore SINGAPORE 365
George Town MALAYSIA 365
Balikpapan INDONESIA 365

North America

Kingston JAMAICA 362
Santiago de Cuba CUBA 351
Hamilton CANADA 53
Honolulu UNITED STATES 337
San Pedro Sula HONDURAS 364
Port-au-Prince HAITI 352
Santo Domingo Este DOMINICAN REPUBLIC 363
Acapúlco de Juárez MEXICO 358
Villa Nueva GUATEMALA 189
San Salvador EL SALVADOR 339
Managua NICARAGUA 360
San Jose COSTA RICA 352

Australia & Oceania

Suva FIJI 357
Cairns AUSTRALIA 315
Waitakere NEW ZEALAND 57

South America

Maracaibo VENEZUELA 365
Cartagena COLOMBIA 365
Guayaquil ECUADOR 364
Macapa BRAZIL 365
Chiclayo PERU 259
Asunción PARAGUAY 234
Cochabamba BOLIVIA 0.2
Montevideo URUGUAY 108
Concepción CHILE 5
San Miguel de Tucumán ARGENTINA 152

Methodology: We used dew point temperature data from **WeatherSpark.com** to rank 14,657 cities worldwide based on the number of uncomfortably humid days in 2023 with a dew point of 60°F (16°C) or above. Ties were resolved by considering extreme humidity days (75°F/24°C dew point or above).

More than ten years since his last act of fleeing as a boy from the rising waters swallowing their island home in the Maldives, he brings his family here to this cold, remote corner of the world and now struggles against the brutish, unforgiving climate of Siberia. His journey from the tropical Maldives to India, and now to Siberia - the stark land of ice- had left him calloused but profoundly conscious of the vulnerability of human survival.

It was a long and arduous journey to Siberia for Areef and Laila, full of uncertainty over leaving everything they knew. They went to Siberia with a few packages and the hope that they would not go wrong at this point in life. But Siberia proved to be no haven, either. It did not take much time before it dawned on them that survival could not have been much harder here, far away from the scorching heat. The biting cold felt extreme like the heat they had experienced in the past, and their bodies, used to warmth, fought against Siberia's fierce winters.

Sadly, Laila deteriorated even more under Siberian climatic conditions. The ruthlessness of colds had too much to ask for her fragile state, and she died a short time after her arrival. Rahim and his sister Aisha buried her in frozen grounds, grieving over the toll each relocation was taking on their family. Nonetheless, Rahim knew that he needed to stay strong for his father and sister. Siberia had been a final resort, but now it also appeared just as cruel and unyielding as the lands they'd left.

And this is the story of all the survivors who ran into Siberia hoping that the freezing climate would be their protection, yet there too, they felt the influence of the growing global temperatures. Summers are no longer dry but blisteringly hot as they should be and cause drowsiness and thirst. The winters, although biting and relentless, were no longer a safe haven. Neither did the poles, so out of the way and inhospitable that few humans had ever set foot there, guarantee a port in a storm; the ice was breaking up, storms became more savage, and available food was scarce.

Desperation led some to talk of heading even farther north. Rahim had heard whispers about Greenland, Iceland, and the remaining uninhabited regions of northern Canada. The cruel irony was that the world's chilliest, most desolate regions were now their most precious havens, while the extreme climates remained beyond the limits of most people to survive.

Then there was the Collapse.

It was quiet at first, with whispers and rumors of cities becoming overrun with chaos, food supply chains collapsing, energy grids failing, and governments losing their grip. As climate disasters increased, the brutal truth became clear: that humanity's infrastructure was crumbling due to environmental catastrophes. The elite, with their wealth and foresight, retreated into bunkers, believing they could ride out the worst part of it in extravagance while the rest of the world was left to face the stark reality of survival, learned quickly that their hidden well stocked bunkers were neither hidden nor well stocked. In short order their imagined safety and comfort melted into dens of death and carnage as the contractors and construction workers that built the bunkers opened them like tin cans and extracted the very needs for their own survival. The delusions of "it will not happen to me" survived only until the bunker doors were locked. The wise knew that humanity either survives or dies together. The choice is yours.

It did not come to Siberia like it did to other parts of the world. There was no violence as there was in Europe, no shards of breaking glass as there was in India and other parts of the world, but the effects were there. Supply chains carrying in food and good essentials overnight disappeared. Government support meager a thing to begin with -evaporated. People were on their own, relying on meager local resources.

As the first hundred days of collapse began to unfold, Rahim and the tiny community he found himself a part of came together, combining their resources and knowledge. The collapse was much harder on parts of the world that relied so heavily on global trade, but Siberia's isolation ironically gave it a kind of strength in resilience. Communities accustomed to rough conditions adapted instantly to the local means of survival.

His father was old and frail, but he did what he could. But the weather would not give up on him. Day after day, it was a fight just to stay warm enough and fed well. Rahim and Aisha joined others scouring for anything edible, hunting the occasional small game, trying to plant whatever they could in small greenhouses against the limited seasons. For the first time, Rahim came to realize how tenaciously his family had survived since leaving the Maldives.

As human beings tried in vain to adapt, so also did the other living things in their battle. However, only a few species could bear the brutal conditions brought about by global warming and high humidity. What remained of the once lush and green forests with their vast diversity were hardy shrubs and drought-resistant plants. Most large mammals disappeared into thin air as their habitats vanished, leaving them to be sealed by relentless drought and famine.

In Siberia, Rahim also observed that the local wildlife had changed. Wolves bears and deer became less common; their numbers went down as ecosystems failed. Meanwhile, smaller and more adaptable species like rodents and insects thrived. Hardy plants, shrubby scrub, and heat-resistant weeds poured into the open spaces. Nature, it seemed, was rebuilding in ways suited to this altered climate, even if humanity was losing ground.

The rich parts of the world crumbled or retreated into bunkers, and Siberians endured. They had no choice but to adapt to their new world. The collapse had been so bad that survival skills became an all-important virtue. Foraging, hunting, and trading had become the adopted style of life in communities brought together in a fragile yet resilient network of mutual support.

Families crowded together in makeshift shelters, hoping for warmth. Food was running low, and people returned to traditional ways of surviving the long winter: digging up roots and preserving everything they could. In this scenario, Rahim's knowledge of environmental sciences becomes very important in that he helped design simple solar collectors and efficient ways of heating, hoping to alleviate the burden of cold.

But in all dimensions of life, the fall was severe. No medical care was at all available, and, short of better treatments by modern medicine, small ailments proved fatal. Many times, Rahim would think about his mother, and what he might have done to keep her safe during that time in India. He had now thought that it was his father's end, and he might lose him too.

As the breakdown lasted beyond a hundred days, there was a cadence of survival. In Siberia, people had bartering and trading, family to family, sharing what they could and relying on the skills and strengths of others. Though his father was not well Rahim continued to teach them both survival skills that he learned when he lived in the Maldives as well as in India, survival skills which had seemed so irrelevant in these icy lands of Siberia. Now, though, they had new importance.

Sources of food were scarce but Siberia's natural resources offered small yet crucial openings. Thawed lakes and lake shores were now a source of freshwater fish, and forest edges supplied the yearly berries. Rahim's community learned to create storage pits that insulate food in an outside pit, thus allowing them to be maintained in the deep cold of those seasons. Every resource is a prize hard-won in an impersonal world that reduces life to its most stripped-down needs.

He bonded with his sister, Aisha, as they fought hard to keep their father alive while administering herbs and remedies they learned from other survivors. He made him grow strong, survive, and find meaning in a world that was changed beyond recognition.

With each passing day, Rahim started to feel rays of hope. His studies, once singular in their focus to understand climate changed from an academic distance, became his tool for survival. He learned to live with nature rather than against it and hence began to understand a place for humanity within the limits of earth's ground.

Slowly, the Siberians evolved to become a people bonded together by the cruel conditions that ensured survival and the experiences they learned from the very ground they stood upon. There were no more nations, governments, or economies to coddle them away from the grim realities of their lives- only people, tied together in a shared will to survive. And amid the intense cold, living resources struggle, Rahim saw the rise to what might well survive even after the old world had toppled; a new kind of society.

And so the seasons passed and his father grew better. Rahim couldn't help but wonder if, in some twisted sense, they were living through the birth of a new humanity; one that might just survive the fall and would, somehow eventually, fashion a world to remember not to repeat the mistakes of the dead.

Chapter XII

Earth's Reclamation and Humanity's Fragile Future

Here, the last specks of the human species clung to the edge of the Earth, wallowing in a relatively temperate climate at the polar extremes. Here, with strenuous effort, very limited resources could be drawn from such an unforgiving environment. In these minute communities, they were as far removed from the civilization that sought to conquer continents, oceans, and even the stars. Even the capacity to survive was doubtful now since resources only worsened each subsequent year. There were no cities or towns, just isolated pockets of people pulled together, not by hope, but only by the intransigent instinct to survive.

It was a shared reality for them as they faced every day, there was no turning back. They had heard tales of the days of human glory when cities almost reached up to the clouds and machines could carry people halfway across the globe in hours. But those stories were much older than the oldest myth to them; the world was buried under its own weight of mythologies.

The last people knew that their life was no longer to preserve human achievement but to witness the gradual and silent recapture of the Earth. Trees had begun sprouting themselves, in once emptied and stark places. Where farmlands and cities sprawled, meadows and forests were slowly taking over and animals, long suppressed by concrete and industrial noise, roamed the lands anew. These survivors witnessed this rebirth within the landscape they experienced and how nature, brutally ravaged and killed, will yet thrive again.

In the loneliness of their daily lives, they saw how entire regions they could no longer inhabit thrived as ecosystems. For seventy years human beings assumed that nature was delicate and should be protected and therefore managed. But nature had never needed protection from the Earth itself but from humans who, for the sake of convenience, had once disregarded its balance. And now that this pressure was gone, it recovered at breathtaking speed: forests reclaimed plains; rivers cut new paths; and species long considered extinct reappeared, finding niches in lands that had once been concrete and steel.

Legacy had become a far cry for these last humans. They no longer built monuments or dreamt of inventions; they simply recorded their lives and thoughts in a minor, humble way as if speaking in a hushed tone to some unseen audience some millennia away. And so they recorded their fight for survival, tenacious nature's bent, the final acceptance of their fate. Some wrote letters to future generations that they would probably never read, speaking of both human achievement and mistakes. Others painted or sketched the landscapes they worked in, catching the contrast between the untamed wilderness and their shrunken community. Over time, the survivors tended less to strive for preserving civilization and more to the simple beauty that had overtaken them.

Those scientists and engineers now developed small garden cultivation, gaining wisdom from the pattern in nature rather than forcing their will on it. Old men, who remembered the great cities, described them instead as ancient myths, telling stories of towers poking into the sky and full cars on the roads. The newer generations listened, but it seemed that these stories were myths, almost like one was hearing tales of a lost mystical land. Fewer people were keeping the tradition fresh, and a cultural memory about human beings was gradually disappearing.

Languages began to change or disappear entirely, and old national identities were all but forgotten. They retreated into the immediacy of surroundings, and let their days schooled by the rhythms of nature rather than the mechanical pulse of cities. It was as if, without the burdens of the world they came from, they gained a deeper, more austere sense of understanding in life. As humanity went, Earth reclaimed it, and regained territories ceded in earlier centuries. The once-throbbing cities were deserted, abandoned to the whim of wind and weather; now only ruins, crumbling slowly to dust. Vines, moss, and wild creatures claimed the skeletons of skyscrapers, factories, and highways. Nature embraced what was left of human ambition not out of malice or a desire for revenge but as part of a cycle that would long outlast them.

By the final years, few were left who could remember a time when the collapse had not yet occurred.

The survivors were now fully part of the world they once sought to dominate, not its rulers or stewards but simply one of many species finding ways to exist within it. They had come to understand that their survival was no longer a victory over nature but a humble coexistence. Their aspiration, is no longer the conquest of more acquisition but understanding and simplicity in the minutest things - the growth of a seedling, the call of a bird, and the soft rustle of the wind. Then, finally, the Earth did not mourn the disappearance of humankind.

It continued, as it was forever; changed and expanding, even in patterns too grand for a human lifetime to imagine. But in these last humans, there was something of serenity in the understanding. They were not the purpose of life but rather a pawn within its vast tapestry. The history of humanity was not the end but just one page within an old and infinite book. And so, as the last remnants of human life receded into darkness, the Earth endured. Nature continued quietly at its work in renewal and growth, awaiting what life might be brought back by. And in that still, fragile peace, humanity's last chapter was penned; not in grand monuments or stunning feats but in the gentle lapping, as it were, of the world to itself.

Here came this new beginning, but no more for humanity as it once was but rather for a world that survived us to continue, with or without our presence.

References

1- U.S. Geological Survey (2023). Water-level and recoverable water in storage changes, High Plains Aquifer, predevelopment to 2019 and 2017 to 2019. USGS Scientific Investigations Report 2023-5143. Available at: https://pubs.usgs.gov/sir/2023/5143/ (Accessed: 18 August 2024).

2- UNICEF. (2021). *Water scarcity: The biggest threat to children*. Available at: https://www.unicef.org/reports/water-scarcity (Accessed: 18 August 2024).

3- World Resources Institute (2023). Water stress in Southern Europe: Current conditions and future projections. Available at: https://www.wri.org/resources/data-visualizations/water-stress-southern-europe (Accessed: 18 August 2024).

4- European Commission (2024) 'Creating a new generation of climate-resilient crops', CORDIS: EU Research Results. Available at: https://cordis.europa.eu/article/id/442178-creating-a-new-generation-of-climate-resilient-crops

5- Wikipedia (2024) 'Farmers' suicides in the United States', Wikipedia. Available at: https://en.wikipedia.org/wiki/Farmers%27_suicides_in_the_United_States

6- Illustrative examples of individual monitoring wells that record cases for which,' ResearchGate. Available at: https://www.researchgate.net/figure/Illustrative-examples-of-individual-monitoring-wells-that-record-cases-for-which_fig2_377663334

7- Climate.gov, 2023. Climate Change: Global Sea Level. [online] Available at: https://www.climate.gov/news-features/understanding-climate/climate-change-global-sea-level [Accessed 4 September 2024].

8- Earth.org, 2023. Sea Level Rise by 2100: Maldives. [online] Available at: https://earth.org/data_visualization/sea-level-rise-by-2100-maldives/ [Accessed 4 September 2024].

9- The Human Climate Fund (HCF), 2024. How do greenhouse gases cause global warming? [online] Available at: https://www.thehcf.org/report-3-how-do-greenhouse-gases-cause-global-warming

10- United Nations, 2024. What is climate change? [online] Available at: https://www.un.org/en/climatechange/what-is-climate-change

11- NOAA National Centers for Environmental Information, 2024. *Global Climate Report - July 2024.* [online] Available at: https://www.ncei.noaa.gov/access/monitoring/monthly-report/global/202407 [Accessed 4 September 2024].

12- The Climate Reality Project. (2023) *How feedback loops are making the climate crisis worse.* Available at: https://www.climaterealityproject.org/blog/how-feedback-loops-are-making-climate-crisis-worse

13- https://carboncredits.com/how-direct-air-capture-works-and-4-important-things-about-it/

14- https://www.bbc.com/future/article/20210310-the-trillion-dollar-plan-to-capture-co2

15- Yoshida, S., Hiraga, K., Takehana, T., Taniguchi, I., Yamaji, H., Maeda, Y., Toyohara, K., Miyamoto, K., Kimura, Y., Oda, K., 2016. A bacterium that degrades and assimilates poly(ethylene terephthalate). *Science,* 351(6278), pp.1196-1199.

16- Eurofish, 2023. *Can plastic-eating bacteria offer a viable solution to the problem of oceanic waste?* [online] Available at: <https://eurofish.dk/can-plastic-eating-bacteria-offer-a-viable-solution-to-the-problem-of-oceanic-waste/>.

17- Yahoo Finance, 2024. *Big tech investments reignite debate over advanced nuclear reactors.* [online] Available at: <https://finance.yahoo.com/news/big-tech-investments-reignite-debate-over-advanced-nuclear-reactors-133016399.html> [Accessed 30 October 2024].

18- Mother Jones. (2022). *The fossil fuel industry's disastrous plans for future oil and gas exploration.* Retrieved from https://www.motherjones.com/politics/2022/05/guardian-investigation-fossil-fuel-oil-gas-industry-plans-exploration-climate-disaster/

19- The Guardian (2024) *Big oil firms knew of dire effects of fossil fuels as early as 1950s, memos show.* Available at: https://www.theguardian.com/us-news/2024/nov/12/big-oil-fossil-fuel-warning?CMP=oth_b-aplnews_d-1

20- https://www.sciencefocus.com/planet-earth/100-humidity-heatwaves-are-spreading-across-the-earth-thats-a-deadly-problem-for-us

21- https://housefresh.com/the-most-humid-cities-in-the-world/